CHEMISTRY RESEARCH AND APPLICATIONS

CHEMISTRY RESEARCHER BIOGRAPHICAL SKETCHES AND RESEARCH SUMMARIES

Chemistry Research and Applications

Additional books in this series can be found on Nova's website
under the Series tab.

Additional e-books in this series can be found on Nova's website
under the e-book tab.

CHEMISTRY RESEARCH AND APPLICATIONS

CHEMISTRY RESEARCHER BIOGRAPHICAL SKETCHES AND RESEARCH SUMMARIES

ELNUR S. HÜSEYNOV
AND
VUSAL BABAYEV
EDITORS

nova publishers
New York

For permission to use material from this book please contact us:
Telephone 631-231-7269; Fax 631-231-8175
Web Site: http://www.novapublishers.com

Library of Congress Cataloging-in-Publication Data

Chemistry research, biographical sketches, and research summaries / editors, Elnur S. H|seynov and Vusal Babayev.
p. cm.
ISBN 978-1-62100-771-5 (hardcover)
1. Chemistry--Research. 2. Chemistry--Biography. I. H|seynov, Elnur S. II. Babayev, Vusal.
QD40.C4533 2011
540--dc23
2011038322

Published by Nova Science Publishers, Inc. † New York

CONTENTS

Preface		**xi**
Part 1 - Research Biographies		
Chapter 1	Saber Ahmed	**3**
Chapter 2	Nailya S. (Sayfullovna) Akhmadullina	**7**
Chapter 3	Irena S. Akhrem	**9**
Chapter 4	Leon A. Apresyan	**11**
Chapter 5	Jose L. Arias	**13**
Chapter 6	Maria I. Arriortua	**17**
Chapter 7	Bin Bao	**23**
Chapter 8	Bianco-Peled Havazelet	**31**
Chapter 9	Vladimir Burtman	**35**
Chapter 10	Vito Capozzi	**39**
Chapter 11	Igor Čeliković	**53**
Chapter 12	Monica C. Chuong	**55**
Chapter 13	Maria Vittoria Diamanti	**57**
Chapter 14	Sergey V. Dorozhkin	**59**
Chapter 15	Snežana Dragović	**61**
Chapter 16	Richard Drevet	**63**
Chapter 17	Barbora Drtinova	**65**
Chapter 18	Pavel N. D'yachkov	**67**
Chapter 19	Fernando S. García Einschlag	**69**
Chapter 20	Sami Fattouch	**73**
Chapter 21	Kikku Fukushima	**79**

Chapter 22	Alexander Y. Galashev	**83**
Chapter 23	Julie Garden-Robinson	**85**
Chapter 24	Penka Dimitrova Gatseva	**89**
Chapter 25	Gennady Gerasimov	**91**
Chapter 26	Cecilia C. Guedes-Silva	**93**
Chapter 27	Norma Guemes Vera	**95**
Chapter 28	Francisco Javier Guillén Gerada	**97**
Chapter 29	Sergio Zarazua Guzman	**99**
Chapter 30	Eduard Hanslík	**101**
Chapter 31	H. James Harmon	**103**
Chapter 32	Masashi Hatanaka	**105**
Chapter 33	Robert B. Heimann	**107**
Chapter 34	Kazutaka Hirakawa	**109**
Chapter 35	Tomohisa Hirobe	**113**
Chapter 36	Gorazd Hribar	**117**
Chapter 37	Bouraoui Ilahi	**119**
Chapter 38	Piera Iuliani	**123**
Chapter 39	Songsri Kaewsuwan	**125**
Chapter 40	C. Karunakaran	**127**
Chapter 41	Shahed U.M. Khan	**131**
Chapter 42	Renat Khaydarov	**133**
Chapter 43	Masashi Kijima	**137**
Chapter 44	Joong Kyun Kim	**139**
Chapter 45	Jean Le Bras	**141**
Chapter 46	Chu-Hsuan Lin	**143**
Chapter 47	Cheng Liu	**145**
Chapter 48	Xiaoxuan Liu	**149**
Chapter 49	Yulia Sergeevna Lukina	**159**
Chapter 50	Maria Virginia Luna	**161**
Chapter 51	Randall D Maples	**165**
Chapter 52	María de los Ángeles Martín Santos	**167**
Chapter 53	Jyri-Pekka Mikkola	**171**

Chapter 54	Steven J Milne	**177**
Chapter 55	Vinita Mishra	**181**
Chapter 56	Yuichi Morishita	**185**
Chapter 57	Jane Morrell	**187**
Chapter 58	Fathi Moussa	**193**
Chapter 59	N. S. Muravyeva	**195**
Chapter 60	Alexandr P. Novikov	**199**
Chapter 61	Hiroaki Onoda	**201**
Chapter 62	Guangnan Ou	**203**
Chapter 63	Byung-Dae Park	**205**
Chapter 64	Jin Hee Park	**207**
Chapter 65	Sang-Eon Park	**211**
Chapter 66	Bertil Rolf Ragnar Persson	**213**
Chapter 67	Eva Petrovova	**219**
Chapter 68	Matthew Picklo	**221**
Chapter 69	Carolina Henritta Pohl	**227**
Chapter 70	Juan Carlos Poveda Jaramillo	**231**
Chapter 71	Tatiana Prado	**233**
Chapter 72	Antonio Jose Queimada	**237**
Chapter 73	Stéphane L. Raeppel	**241**
Chapter 74	Michael A. Reshchikov	**245**
Chapter 75	Claudionor Ribeiro da Silva	**247**
Chapter 76	Semenov Semen	**249**
Chapter 77	Konstantin Zh.Seminsky	**251**
Chapter 78	Mohd. Shakir Khan	**253**
Chapter 79	Jose Angel Siles Lopez	**255**
Chapter 80	Steven Y. Qian	**259**
Chapter 81	Y. Tai Clifford	**261**
Chapter 82	Tomoki Takahashi	**265**
Chapter 83	Shinji Tokonami	**269**
Chapter 84	Vukosava Milic Torres	**275**
Chapter 90	Inés Velasco	**295**

Chapter 91	Gloria Víllora	**299**
Chapter 92	Maja D. Vitoroviæ-Todoroviæ	**303**
Chapter 93	Vladimir Volodin	**305**
Chapter 94	Winfried Vonau	**309**
Chapter 95	Andriy Voronov	**313**
Chapter 96	Viroj Wiwanitkit	**317**
Chapter 97	Anatoly G. Zakharov	**319**
Chapter 98	Aristides D. Zdetsis	**321**
Chapter 99	Jasenka Piljac Zegarac	**325**
Chapter 100	Zheng-Wen Li	**329**
Chapter 101	Jun Zhu	**333**
Chapter 102	Sergei P.Zhvavyi	**337**
Chapter 103	Evangelos Zoidis	**339**
Chapter 104	Zou Jin-Long	**343**
Part 2 - Research Summaries in Chemistry		
Chapter 105	Advances in Adsorption Technology *Bidyut Baran and Kim Choon Ng*	**347**
Chapter 106	Advances in Chemical Modeling *Mihai V. Putz*	**349**
Chapter 107	Analytical Chemistry of Cadmium: Sample Pre-Treatment and Determination Methods *Antonio Moreda-Piñeiro and Jorge Moreda-Piñeiro*	**351**
Chapter 108	Artificial Neural Network Modeling of Water and Wastewater Treatment Processes *Ali R. Khataee and Masoud B. Kasiri*	**353**
Chapter 109	Biomedical Applications of Lanthanum *Harriet Nilsson, Anca Dragomir and Godfried M. Roomans*	**355**
Chapter 110	Boron Hydrides, High Potential Hydrogen Storage Materials *Umit B. Demirci and Philippe Miele*	**357**
Chapter 111	Chemistry and Biochemistry: From Pure to Applied Science, New Horizons-Vol. 3 (A Festschrift in Honor of the 75th Birthday of Professor Gennady E. Zaikov) *Eli M. Pearce and G. E. Zaikov*	**359**
Chapter 112	Chemistry and Chemical Engineering Research Progress *A. K. Haghi*	**361**

Chapter 113 Chitosan-Based Hydrogels for Tissue Engineering Applications **363**
G. Luna-Bárcenas, E. Prokhorov, E. Elizalde-Peña, A. Nuno-Licona, I. C. Sanchez, J. E. Gough, C. Velasquillo-Martinez and C. E. Schmidt

Chapter 114 Cyclic B-Ketoesters: Synthesis and Reactions **365**
M. A. Metwally and E. G. Sadek

Chapter 115 Dzhemilev Reaction in Organic and Organometallic Synthesis **367**
Vladimir A. D'yakonov

Chapter 116 Electrochemical Hydrogen and Metals: Absorption, Diffusion and Embrittlement Prevention in Corrosion and Electroplating **369**
Sergei Beloglazov

Chapter 117 Electrochemical Properties and Applications of Ionic Liquids **371**
Angel A. J. Torriero and Muhammad J. A. Shiddiky

Chapter 118 Electrospray Technology for Thin-Film Deposition **373**
Anatol Jaworek

Chapter 119 Emerging Topics in Organic Chemistry **375**
Rattaporn Thonggom, Thevarak Rochanapruk and Gennady E. Zaikov

Chapter 120 Humic Acids from Raw Materials of the Czech Republic **377**
Libuse Madronova

Chapter 121 Information Origins of the Chemical Bond **379**
Roman F. Nalewajski

Chapter 122 Layered Double Hydroxides: Present and Future **381**
Vicente Rives

Chapter 123 Magnetic Interactions in Oxo-Carboxylate Bridged Gadolinium (III) Complexes **383**
Laura Cañadillas-Delgado, Óscar Fabelo, Catalina Ruiz-Pérez and Joan Cano

Chapter 124 Molecular and Supramolecular Bioinorganic Chemistry: Applications in Medical and Environmental Sciences. Volume 2 **385**
Ana Lucia Ramalho Mercê, Judith Felcman and María Ángeles Lobo Recio

Chapter 125 Molecular Symmetry and Fuzzy Symmetry **387**
Xuezhuang Zhao

Chapter 126 Nanotube Array Supercapacitor **389**
Yibing Xie

Chapter 127 Optimization in Polymer Processing **391**
António Gaspar-Cunha and José António Covas

Chapter 128 Organic Europium Complexes and their Applications in Optoelectronic Devices **393**
Han You and Dongge Ma

Chapter 129 Phosphazenes: A Worldwide Insight **395**
R. De Jaeger and M. Gleria

Chapter 130 Physical Chemistry Examinations **397**
A. O. Ibhadon

Chapter 131 Polymer and Composites: Theory and Practical Applications **399**
G. V. Kozlov, A. K. Mikitaev and G. E. Zaikov

Chapter 132 Preparation of Thin Film Pd Membranes for H2 Separation From Synthesis Gas and Detailed Design of a Permeability Testing Unit **401**
M. Bientinesi and L. Petarca

PREFACE

This new book compiles biographical sketches of top professionals in the field of chemistry, as well as research summaries from a number of different focuses in this important field.

PART 1 - RESEARCH BIOGRAPHIES

In: Chemistry Researcher Biographical Sketches
Editors: E. S. Huseynov and V. Babayev
ISBN: 978-1-62100-771-5

Chapter 1

BIOGRAPHICAL SKETCH

Name:	Title:	Date of Birth:
Saber Ahmed	Mr	16.12. 1975

EDUCATION:

Institution and Location	Degree	Year Conferred	Scientific Field
M.Eng.-Urban and Environmental Engineering, 2002, HOKKAIDO UNIVERSITY, Japan	PhD Candidate, CQ University, Australia	2011	
BSc. Eng.-Civil 1999, Bangladesh University of Engineering & Technology (BUET)			

CONTACT POINTS:

Building 29/1.03, Faculty of SEH, CQ University, North rockhampton, QLD-4702, Australia, 617 049309634 (Work), Email: s.ahmed@cqu.edu.au

RESEARCH AND PROFESSIONAL EXPERIENCE:

- Water and wastewater treatment processes
- Photocatalytic removal of organic dyes and phenolic contaminants from wastewater
- Photocatalytic removal of pesticides and herbicides from wastewater
- Membrane and chemical treatment of water and wastewater

- Arsenic and humic substances removal from drinking and wastewater using membrane filtration
- Pilot testing of membrane and chemical treatment systems
- Drinking water and wastewater treatment system planning
- Physical and morphological modelling of river to support the design of river structure
- Competent in Software: SPSS, GAMBIT, Fluent 6.26, Auto CAD

Professional Appointments:

EMPLOYER: CENTRAL QUEENSLAND UNIVERSITY, AUSTRALIA,
Position: Research Assistant (From August 2008 to Present)
EMPLOYER: RAJDHANI UNNAYAN KARTRIPAKKHA (RAJUK) Bangladesh, Dhaka-1000,
Position : Senior Assistant Engineer-Civil (From Nov. 2003 to July 2008)
EMPLOYER: RIVER RESEARCH INSTITUTE, Hydraulic Research Directorate, Bangladesh
Position : Scientific Officer (May 2000 to Oct. 2003)
EMPLOYER : SAIDABAD SURFACE WATER TREATMENT PLANT PROJECT, Dhaka Water Supply and Sewerage Authority, Bangladesh
Position : Consultant Engineer (November1999 to May 2000)

Publications during last three years

- Saber Ahmed, M.G. Rasul, M.A. Hashib, Y. Watanabe. Performance of Nanofiltration membrane in a Vibrating module (VSEP-NF) for arsenic removal. Desalination 252, 2010, 127- 134
- Saber Ahmed, M.G. Rasul, W.N. Martens, R. Brown, M. A. Hashib. Heterogeneous photocatalytic degradation of phenols in wastewater: A review on current status and developments. Desalination 261(1-2),3-18
- Saber Ahmed, M.G. Rasul, W.N. Marten, R. Brown., M.A. Hashib. Advances in Heterogeneous Photocatalytic Degradation of Phenols and Dyes in Wastewater: A Review. Water, Air, & Soil Pollution DOI 10.1007/s11270-010-0456-3(Springer)
- Sang-Hyup Yi, Saber Ahmed, Watari, Yoshimasa Watanabe. Arsenic removal by MF membrane with chemical sludge adsorption and NF membrane equipped with vibratory shear enhanced process (VSEP), Water Science and Technology: Water Supply (IWA), Vol.3, 2003, 303-310
- Saber Ahmed, M.G. Rasul, R. Brown., and M.A. Hashib. Influence of Parameters on the Hetererogeneous Photocatalytic Degradation of Pesticides and Phenolic Contaminants in Wastewater: A short review. Journal of Environmental Management 92(3), 2011,311-33

- Saber Ahmed,S., Seraji,T.M., Jahedi,J., Hashib,M.A. CFD simulation of turbulence promoters in a tubular membrane channel. Desalination 276 ,2011,191–198

Conference Publications

- Saber Ahmed, M.G. Rasul, W. Marten, R. Brown Computational fluid dynamics analysis of a flat plate photocatalytic reactor for storm and wastewater reuse. Proceedings of the 13th Asian Congress of Fluid Mechanics, 17-21 December 2010, Dhaka
- Saber Ahmed, M.G. Rasul, W. Marten, R. Brown. Sustainable Reuse of Storm and Waste-water Resources: A review on Solar Photocatalytic Water Purification Processes. Presented at the conference on "Regional Sustainability: Local solution to Global issues" of CREC 2009, 14-15 August 2009, CQ university, Rockhampton, Australia
- Saber Ahmed, M.G. Rasul, W. Marten, R. Brown., A review on the role of operating parameters in heterogeneous photocatalytic water purification process for storm and waste water reuse. Proceedings of the 5th WESEAS International conference on " Recent Advances in Energy and Environment " University of Cambridge, UK ,20-25 February, 2010, ISBN: 978-960-474-159-5
- Saber Ahmed, M.G.Rasul. Wayde N.Marten, R. Brown. Recent developments in photocatalytic degradation of pesticides and phenols in storm and wastewater effluent. Proceedings of the Environmental Research Event ,ERE 2010, CQ University, Rockhampton, QLD
- Sang-Hyup Yi, Saber Ahmed, Watari, Yoshimasa Watanabe. Arsenic Removal by MF membrane with Chemical Sludge Adsorption and NF membrane Equipped With Vibratory Shear Enhanced Process (VSEP), Published in the proceedings of the International Conference on" Membranes in Drinking Water and Industrial Water Production (MDIW), September 2002, Mulheim an der Ruhr, Germany.

In: Chemistry Researcher Biographical Sketches
Editors: E. S. Huseynov and V. Babayev
ISBN: 978-1-62100-771-5

Chapter 2

BIOGRAPHICAL SKETCH

Name:	Title:	Date of Birth:
Nailya S. (Sayfullovna) Akhmadullina	PhD	06-03-1984

EDUCATION:

Institution and Location	Degree	Year Conferred	Scientific Field
Mari State University, 424001, Russia, Republic of Mari El, Yoshkar-Ola, Lenin square, 1	PhD		
A.A. Baikov Institute of metallurgy and material science of Russian Academy of Sciences			

CONTACT POINTS:

119991, Russia, Moscow, Leninsky Avenue, 49

RESEARCH AND PROFESSIONAL EXPERIENCE:

2006 PhD-student in N.S. Kurnakov Institutte of General and Inorganic Chemistry of Russian Academy of Sciences,
2011 Researcher in A.A. Baikov Instiitute of metallurgy and material science of Russian Academy of Sciences.

Professional Appointments:

Noble metals chemistry

Honors: Laureate of the Chernyaevâ's Prize

Publications during last three years:

1. N.S.Akhmadullina, N.V. Cherkashina, M.N. Vargaftik, I.I. Moiseev et al., «Synthesis of palladium(II)3d-metal(II) paddlewheel acetate-bridged heterodimetallic complexes: Unexpected catalysis by water molecules» // Inorg. Chim. Acta, 2009, V. 362, P. 1943.
2. A. A. Markov, A. P. Klyagina, S. P. Dolin, N. S. Akhmadullina, N. Yu. Kozitsyna, N. V. Cherkashina, S. E. Nefedov, M. N. Vargaftik, I. I. Moiseev, «Chemical bonding in heterobimetallic palladium(II) complexes with 3d metals». // Russ. J. Inorg. Chem., 2009, V. 6, P. 950.
3. N.S.Akhmadullina, N.V. Cherkashina, N. Yu. Kozitsyna, A.E. Gekhman, M.N. Vargaftik, «Unexpected participation of nucleophiles in the reaction of palladium(II) acetate with divalent 3d metals». // Kinetics and Catalysis, 2009, V. 3, P. 396.
4. S.E. Nefedov, N.Yu. Kozitsyna, N.S. Akhmadullina, N.V. Cherkashina, M.N. Vargaftik, I.I. Moiseev, «Unusual heterobimetallic palladium(II) cerium(III) acetate complexes». /// Inorg. Chem. Comm., 2011, V. 14, P. 554.
5. S.E. Nefedov, N.Yu. Kozitsyna, N.S. Akhmadullina, N.V. Cherkashina, M.N. Vargaftik, I.I. Moiseev. «The role of water molecules in formation of heterometallic palladium acetate complexes with cerium and neodymium». // Russ. J. Inorg. Chem., 2011, V. 3, P. 399.

In: Chemistry Researcher Biographical Sketches
Editors: E. S. Huseynov and V. Babayev

ISBN: 978-1-62100-771-5

Chapter 3

BIOGRAPHICAL SKETCH

Name:	Title:	Date of Birth:
Irena S. Akhrem	A.N. Nesmeyanov Institute of Organoelement Compounds, Russian Academy of Sciences, Russian Federation, Moscow	01-10-1932

EDUCATION:

Institution and Location	Degree	Year Conferred	Scientific Field
Graduated from Moscow State University		1955	

CONTACT POINTS:

119454, Moscow, Vernagskii pr., 38 ''A", 107

RESEARCH AND PROFESSIONAL EXPERIENCE:

Activation of Si – H and S – C связей in silaparaffines under the action of transition metal complexes and ionic fluorides. Elaboration of effective catalysts for hydrocracking of alkanes and cycloalkanes based on Ziegler – Natta complexes and ligand deficient osmium clusters. Development of superelectrophilic systems of new types for activation and low-temperature transformations of alkanes and cycloalkanes. Hydrocarbon transformations of alkanes and cycloalkanes. Elaboration of selective syntheses of organic compounds from alkanes and cycloalkanes. The nature of new active superelectrophilic systems and mechanistic studies on superelectrophilic alkane activation. "Alkane-like" functionalizations of monofunctional organic compounds. About 250 publications.

Professional Appointments:

Chief of the group of activation of inert compounds

Honors Relevant to Nanomaterials Research:

Honors: Ph. D. (1961), D. Sc. (1991), Professor (1991)

Publications during last three years

1. S. Akhrem, A. V. Orlinkov, Polyhalomethanes Combined with Lewis Acids in Alkane Chemistry, *Chem. Revs.* 2007, *107,* 2037- 2079.
2. Irena Akhrem, Dzhul'etta Avetisyan, Lyudmila Afanas'eva, Sergei Vitt, Pavel Petrovskii, Alexander V. Orlinkov, The first one-pot amidation of alkanes cycloalkanes, *Tetrahedron Letters,* 2008, *49*, 1399 – 1404.
3. Alexander Orlinkov, Nikolai Kagramanov, Pavel. Petrovskii, Irena Akhrem, The first "alkane-like" functionalization of *n*-alkyl acetates: a new method for one-pot selective syntheses of bifunctional aliphatic compounds with an acetate group, *Tetrahedron Letters*, 2010, *51*, 259-263.
4. Irena Akhrem, Dzhul'etta Avetisyan, Lyudmila Afanas'eva, Nikolai Kagramanov, Pavel Petrovskii, Alexander Orlinkov, The first one-pot "alkane-like" reactions of carbonyl-containing adamantanes, Tetrahedron Letters, 2010, 51, 907-911.
5. Irena S. Akhrem, Dzhu'letta V. Avetisyan, Nicolai D. Kagramanov, Pavel V. Petrovskii, Nadezhda E. Mysova, One-pot N-acylation and alkylation of o-nitroaniline by saturated hydrocarbons in the presence of carbon monoxide, *Mendeleev Commun.* 2010, *20*, 5, 257-259.
6. I. S. Akhrem, L.V. Afanas'eva, S. V. Vitt, N. D. Kagramanov P. V. Petrovskii, A. V. Orlinkov, I. M. Churilova. 1,3,5,7 – tetraalkyladamantanes as new adamantane derivatives. The method of preparation of 1,3,5,7 – tetraalkyladamantanes and modification of mixtures of polyalkyladamantanes (RU 2409545, 2008).
7. А. Л. Чистяков. И. В.Станкевич, Н. П. Гамбарян, И.С. Ахрем, Квантово-химическое изучение механизма нитрозирования метана под действием NO и суперэлектрофилов, *ДАН*, 2007, 414, 4, 505 - 508.
8. Irena Akhrem, Dzhul'etta Avetisyan, Lyudmila Afanas'eva. Evgenii Goryunov, Irina Churilova, Pavel Petrovskii and Nikolai Kagramanov *Mendeleev Commun.* 2011 (in press).

In: Chemistry Researcher Biographical Sketches
Editors: E. S. Huseynov and V. Babayev

ISBN: 978-1-62100-771-5

Chapter 4

BIOGRAPHICAL SKETCH

Name:	Title:	Date of Birth:
Leon A. Apresyan	PhD	01.03.1950

EDUCATION:

Institution and Location	Degree	Year Conferred	Scientific Field
Moscow, Physical-Technical Institute (MFTI), Faculty of General and Applied Physics;		1966-1972	

CONTACT POINTS:

Tel. +7(499)132-6868, E-mail: leon_apresyan@mail.ru

RESEARCH AND PROFESSIONAL EXPERIENCE:

Wave propagation in random media, radiation transfer theory, influence of atmosphere and ionosphere on radar and lidar systems, electrodynamics of composite and polymer materials, synthesis of carbon nanostructures.

Professional Appointments:

1972-2001 - Senior Scientist, Mintz Radio Engineering Institute
2001-present - Head of Laboratory, Prokhorov Institute of General Physics

Publications during last three years

1. D. V. Vlasov, L. A. Apresyan, T. V. Vlasova and V. I. Kryshtob. *Nonlinear response and two stable electroconducting states in transparent plasticized PVC films.* Technical Physics Letters, 2010, Volume 36, Number 10, Pages 923-925.
2. L. A. Apresyan and D. V. Vlasov. *On the description of internal impedance per unit length for multi-walled carbon nanotube.* Electromagnetic waves and electronic systems. 2010. 1. (in Russian).
3. L. A. Apresyan, D. V. Vlasov, T. V. Vlasova and A. A. Klimanov. *Experimental study of the influence of hydrogenation on synthesis of carbon nanostructures in a reactor with activated hydrogen.* Technical Physics, 2009, Volume 54, Number 3, Pages 419-422.
4. L. A. Apresyan and D. V. Vlasov *Leontovich-Levin equation taking into account finite conductivity of a conductor.* Technical Physics, 2009, Volume 54, Number 12, Pages 1736-1741.
5. L. A. Apresyan, D. V. Vlasov, V. I. Konov and V. I. Kryshtob. *Catalytic synthesis of tubular carbon nanofibers using propagating combustion wave in acetylene.* Technical Physics Letters, 2009, Volume 35, Number 11, Pages 1065-1067.
6. L. A. Apresyan and D. V. Vlasov. *Polarization and antennas effects in light scattering on elongated anisotropic nano-particles.* Electromagnetic waves and electronic systems. 2008. Volume 13. 9. Pages 56-64 (in Russian).
7. Apresyan L.A., Vlasov D.V., Konov V.I *Activated-Hydrogen Reactor for Synthesis of Carbon Nanotubes..* Carbon 2007. International Conference, Seattle, USA, Juli 15-20, 2007.

In: Chemistry Researcher Biographical Sketches
Editors: E. S. Huseynov and V. Babayev

ISBN: 978-1-62100-771-5

Chapter 5

BIOGRAPHICAL SKETCH

Name:	Title:	Date of Birth:
Jose L. Arias	Department of Pharmacy and Pharmaceutical Technology, School of Pharmacy, University of Granada, Spain	14-06-1976

EDUCATION:

Institution and Location	Degree	Year Conferred	Scientific Field
Degree in Pharmacy, University of Granada		1999.	
Advanced Study Diploma in Physics. University of Granada,		2001	
Advanced Study Diploma in Pharmacology and Industrial Pharmacy. University of Granada		2002.	
Ph.D. from the University of Granada Dissertation: Evaluation of poly (ethyl-2-cyanoacrylate) nanoparticles with a magnetic core as systems Cumlaude for the controlled release of the antitumour drug 5-fluorouracyl.		(2003).	

CONTACT POINTS:

Department of Pharmacy and Pharmaceutical Technology, School of Pharmacy, University of Granada. Campus Universitario de Cartuja s/n, 18071 Granada, Spain.

RESEARCH AND PROFESSIONAL EXPERIENCE:

Stimuli-sensitive drug nanocarriers.
Long circulating nanoparticles. Nano-oncology. Nanomedicines against severe diseases, Multifunctional nanoplatforms for biomedical applications.
Nanomedicines against severe diseases.

Professional Appointments:

November 2005-present - Assistant professor at the Department of Pharmacy and Pharmaceutical Technology, University of Granada.

Publications during last three years

- José L. Arias, Amparo Gómez-Gallo, Angel V. Delgado, Visitación Gallardo. Study of the stability of Kollidon SR for pharmaceutical applications. Colloids and Surfaces A: Physicochemical and Engineering Aspects 338, 107-113 (2009).
- José L. Arias, Amparo Gómez-Gallo, Angel V. Delgado, Mª. Adolfina Ruiz. Kollidon SR colloidal particles as vehicles for oral morphine delivery in pain treatment. Colloids and Surfaces B: Biointerfaces 70, 207-212 (2009).
- M.A. Holgado, M.J. Cózar-Bernal, S. Salas, J.L. Arias, J. Alvarez-Fuentes, M. Fernández-Arévalo. Protein-loaded PLGA microparticles engineered by flow focusing: Physicochemical characterization and protein detection by reversed-phase HPLC. International Journal of Pharmaceutics 380, 147-154 (2009).
- José L. Arias, L. Harivardhan Reddy, Patrick Couvreur. Polymeric nanoparticulate system augmented the anticancer therapeutic efficacy of gemcitabine. Journal of Drug Targeting 17, 586-598 (2009).
- José L. Arias, Margarita López-Viota, Julian López-Viota, Angel V. Delgado. Development of iron/ethylcellulose (core/shell) nanoparticles loaded with diclofenac sodium for arthritis treatment. International Journal of Pharmaceutics 382, 270-276 (2009).
- José L. Arias, Margarita López-Viota, Eva Sanchez-Fernandez, Mª. Adolfina Ruiz. Formulation and physicochemical characterization of poly (Îµ-caprolactone) nanoparticles loaded with ftorafur and diclofenac sodium. Colloids and Surfaces B: Biointerfaces 75, 204-208 (2010).

- José L. Arias, Gema I. MartÍnez-Soler, Margarita López-Viota, Mª. Adolfina Ruiz. Formulation of chitosan nanoparticles loaded with metronidazole for the treatment of infectious diseases. Letters in Drug Design & Discovery 7, 70-78 (2010).
- José L. Arias, Margarita López-Viota, Angel V. Delgado, Mª. Adolfina Ruiz. Iron/ethylcellulose (core/shell) nanoplatform loaded with 5-fluorouracil for cancer targeting. Colloids and Surfaces B: Biointerfaces 77, 111-116 (2010).
- Mª José Cózar-Bernal, Visitación Gallardo, Eva Sáez-Fernández, Mª Angeles Holgado, Josefa Alvarez-Fuentes, Mercedes FernÁndez-ArÉvalo,
- José L. Arias. Role of the electrokinetic properties on the stability of mebendazole suspensions for veterinary applications. International Journal of Pharmaceutics 393, 161-166 (2010).
- José L. Arias, Margarita López-Viota, Visitación Gallardo, María Adolfina Ruiz. Chitosan nanoparticles as a new delivery system for the chemotherapy agent tegafur. Drug Development and Industrial Pharmacy 36, 744-750 (2010).
- José L. Arias, L. Harivardhan Reddy, Patrick Couvreur. Superior preclinical efficacy of gemcitabine developed as chitosan nanoparticulate system. Biomacromolecules 12, 97-104 (2011).
- José L. Arias. Drug targeting strategies in cancer treatment: An overview. Mini-Reviews in Medicinal Chemistry 11, 1-17 (2011).
- José L. Arias, L. Harivardhan Reddy, Mohammad Othman, Brigitte Gillet, Didier Desmaí«le, Fatima Zouhiri, Franco Dosio, Ruxandra Gref, Patrick Couvreur. Squalene based nanocomposites: A new platform for the design of multifunctional pharmaceutical theragnostics. ACS Nano 5, 1513-1521 (2011).
- José L. Arias, Margarita López-Viota, Eva Sáez-Fernández, Mª Adolfina Ruiz, Angel V. Delgado. Engineering of an antitumor (core/shell) magnetic nanoformulation based on the chemotherapy agent ftorafur. Colloids and Surfaces A: Physicochemical and Engineering Aspects 384, 157-163 (2011).
- M.A. Holgado, J. Alvarez-Fuentes, M. Fernández-Arévalo, J.L. Arias. Possibilities of poly(D,L-lactide-co-glycolide) in the formulation of nanomedicines against cancer. Current Drug Targets 12, 1096-1111 (2011).
- J.L. Arias, B. Clares, M.E. Morales, V. Gallardo, M.A. Ruiz. Lipid-Based Drug Delivery Systems for Cancer Treatment. Current Drug Targets 12(8), 1151-1165 (2011).

In: Chemistry Researcher Biographical Sketches
Editors: E. S. Huseynov and V. Babayev
ISBN: 978-1-62100-771-5

Chapter 6

BIOGRAPHICAL SKETCH

Name:	Title:	Date of Birth:
Maria I. Arriortua	Universidad del País Vasco/Euskal Herriko Unibertsitatea	11/11/1950

EDUCATION:

Institution and Location	Degree	Year Conferred	Scientific Field
Doctor in Chemical Sciences			

CONTACT POINTS:

Dpto. Mineralogía y Petrología, Apdo. 644, E-48080 Bilbao. Spain

RESEARCH AND PROFESSIONAL EXPERIENCE:

Maria I. Arriortua ended her doctoral thesis at the University of Lovaine-la-Neuve (Belgium), specialized in Solid State Chemistry. Since 1976 she was lecturer of the University of the Basque Country (UPV/EHU) and since 1992 she is Professor of Crystallography and Mineralogy of the UPV/EHU. She has set up a research group, leading several research projects and achieving great scientific equipment (X-ray diffraction, magnetic measurements …) at competition. She haas directed doctoral thesis in the field of Materials Crystallography, published in specialized journals (267) and made communications at national and international congresses (402). H= 33.

Since 1980, she is charge of the X-ray services of the Science and Technology Faculty of the UPV/EHU and since 2002, she is the Head of the Research General Services of the

UPV/EHU. She was also vice-president of the Spanish Crystallography group (1996-2010) and member of the National Committee of the International Union of Crystallography (1985-present). She has take part of several national and international congresses scientific and organizing committees and she had been member of evaluation committees: Comisión Nacional Evaluadora de la Actividad Investigadora CNEAI• , various experts ‘ committees of the Ministry of Education and Science (MEC) for research activities and human resources. She has contributed to the training of 95 researchers in the UPV/EHU. Nowadays, she takes part of the experts committee of the Basque agency UNIQUAL and she is member of the Executive Council of the Management for Internationalization of Innobasque. Recently, she has been nominated as coordinator of Science and Technology Area of the National Evaluation and Foresight Agency, ANEP (since February 2010).

Professional Appointments:

Professor of crystallography and mineralogy

Publications during last three years

- Authors: B. Bazán, J.L. Mesa, J.L. Pizarro, A. Peña, E. Legarra, M.I. Arriortua, T. Rojo.
  Tittle: Structural, Thermal, Spectroscopic and Magnetic Studies of the $(C_3H_{12}N_2)_{0.75}[Mn^{II}{}_{1.5}Fe^{III}{}_{1.5}(AsO_4)F_6]$ and $(C_3H_{12}N_2)_{0.75}[Co^{II}{}_{1.5}Fe^{III}{}_{1.5}(AsO_4)F_6]$ Compounds.
  Mater. Res. Bull., 43, 1307-1320, 2008
- Authors: T. Berrocal, J. L. Mesa, J. L. Pizarro, L. Lezama, B. Bazán, M. I. Arriortua, T. Rojo.
  Tittle: Hydrothermal Synthesis, Crystal Structures, Thermal Behavior and Spectroscopic and Magnetic Properties of Two New Organically Templated Fluoro-Vanadyl-Hydrogenarsenates: $(R)_{0.5}[(VO)(HAsO_4)F]$ (R: Ethylenediammonium and Piperazinium).
  J. Solid State Chem., 181, 884-894, 2008
- Authors: Y. Vara, E. Aldaba, A. Arrieta, J.L. Pizarro, M.I. Arriortua, F.P. Cossío.
  Tittle: Regiochemistry of the Microwave-Assisted Reaction Between Aromatic Amines and ï• ¡-Bromoketones to Yield Substituted 1H-Indoles.
  Org. Biomol. Chem., 6, 1763-1772, 2008
- Authors: A. Larrañaga, J.L. Mesa, J.L. Pizarro, L. Lezama, M.I. Arriortua, T. Rojo.
  Tittle: Synthesis, Thermal, Spectroscopic and Magnetic Studies of the $Mn(SeO_3)\cdot 2H_2O$ and $Fe_2(SeO_3)_3\cdot 3H_2O$ Selenites.
  Spectrochim. Acta A, 69A(3), 1020-1026, 2008
- Authors: B. Inés, R. SanMartín, F. Churruca, E. Domínguez, M.K. Urtiaga, M.I. Arriortua.
  Tittle: A Nonsymmetric Pincer-Type Palladium Catalyst In Suzuki, Sonogashira, and Hiyama Couplings in Neat Water.
  Organometallics, 27, 2833-2839, 2008

- Authors: R. Fernández de Luis, J.L. Mesa, M.K. Urtiaga, L. Lezama, M. I. Arriortua, T. Rojo.
  Tittle: Topological Description of a 3D Self-Catenated Nickel Hybrid Vanadate $Ni(bpe)(VO_3)_2$. Thermal Stability, Spectroscopic and Magnetic Properties.
  New J. Chem., 32,1582-1585, 2008
- Authors: A. Martínez-Amesti, A. Larrañaga, L.M. Rodríguez-Martínez, A.T. Aguayo, J.L. Pizarro, M.L. Nó,
  Laresgoiti, M.I. Arriortua.
  Tittle: Reactivity Between $La(Sr)FeO_3$ Cathode, Doped CeO_2 Interlayer and Yttria-Stabilized Zirconia Electrolyte for Solid Oxide Fuel Cell Applications.
  J. Power Sources, 185(1), 401- 410, 2008.
- Authors: P. Gómez-Saiz, R. Gil-García, M.A. Maestro, J.L. Pizarro, M.I. Arriortua, L. Lezama, T. Rojo, M. González-Alvarez, J. Borrás, J. García-Tojal.
  Tittle: Structure, Magnetic Properties and Nuclease Activity of Pyridine-2-carbaldehyde thiosemicarbazonecopper(II) Complexes.
  J. Inorg. Biochem., 102, 1910-1920, 2008
- Authors: T. Berrocal, J.L. Mesa, J.L. Pizarro, B. Bazán, M. Iglesias, A.T. Aguayo, M.I. Arriortua, T. Rojo.
  Tittle: Microporous Vanadyl-Arsenate with the Template Incorporated Exhibiting Sorption and Catalytic Properties.
  Chem. Commun., 39, 4738-4740, 2008
- Authors: Y. Vara, T. Bello, E. Aldaba, A. Arrieta, J. L. Pizarro, M. I. Arriortua, X. López, F. P. CossìÌ• o.
  Tittle: Trans-Stereoselectivity in the Reaction between Homophthalic Anhydride and Imines.
  Org. Lett, .10(21), 4759-4762, 2008
- Authors: T. Berrocal, J. L. Mesa, J. L. Pizarro, B. Bazán, I. Ruiz de Larramendi, M.I. Arriortua, T. Rojo.
  Tittle: $(NH_4)[V^{III}{}_{1-x}V^{IV}{}_x (AsO_4)F_{1-x}O_x]$: A New Mixed Valence Vanadium(III,IV) Fluoro-Arsenate with Ferromagnetic Interactions and Electronic Conductivity.
  J. Solid State Chem., 182, 65-71, 2009
- Authors: A. Larrañaga, J. L. Mesa, L. Lezama, J. L. Pizarro,, M.I. Arriortua, T. Rojo.
  Tittle: Supercritical Hydrothermal Synthesis of $Cu_2O(SeO_3)$: Structural Characterization, Thermal, Spectroscopic and Magnetic Studies.
  Mater. Res. Bull., 44, 1- 5, 2009
- Authors: P. Gómez-Saiz, R. Gil-García, M. A. Maestro, F. Arnaiz, L. Lezama, T. Rojo,J. L. Pizarro, M. I. Arriortua, L. M. Gonzalez-Alvarez, J. Borrás, V. Diez-Gomez, J. García-Tojal.
  Tittle: 1,3,4-Oxadiazolecopper(II) Compounds: Dimensionality, Magnetism and Nuclease Activity.
  Eur. J. Inorg. Chem., 373-388, 2009
- Authors: A. Larrañaga, J. L. Mesa, L. Lezama, M.I. Arriortua, T. Rojo.
  Tittle: Mild Hydrothermal Synthesis of $Cu_2O(SeO_3)2H_2O$: Structural Characterization, Thermal, Spectroscopic and Magnetic Studies.
  Spectrochimica Acta Part A 72, 356-360, 2009

– Authors: T. Berrocal, J. L. Mesa, J. L. Pizarro, B. Bazán, I. Ruiz de Larramendi, M.I. Arriortua, T. Rojo.
Tittle: Mild Hydrothermal Synthesis, Crystal Structure, Thermal Behaviour, Spectroscopic and Magnetic Properties of $(NH_4)[Fe(AsO_4)_{1-x}(PO_4)_xF]$ (x= 0.3, 0.6, 0.8) Series. Thermal Transformation of $(NH_4)[Fe(AsO_4)_{0.7}(PO_4)_{0.3}]F$ into the Textural Porous Orthorhombic $Fe(AsO_4)_{0.7}(PO_4)_{0.3}$.
J. Solid State Chem., 182, 932-941, 2009
– Authors: T. Rojo, J. L. Mesa, J. Lago, B. Bazán, J. L. Pizarro, M. I. Arriortua.
Tittle: Organically Templated Open-Framework Phosphite.
J. Mater. Chem.19, 3793-3818, 2009
– Authors: R. Fernández de Luis, M.K. Urtiaga, J.L. Mesa, T. Rojo, M. I. Arriortua.
Tittle: Structural Archetypes in Nickel Hybrid Vanadates. Towards a Directed Hydrothermal Synthesis.
J. Alloy. Compd., 480, 54-56, 2009
– Authors: N. Ortiz-Vitoriano, I. Ruiz de Larramendi, J.I. Ruiz de Larramendi, M.I. Arriortua, T. Rojo.
Tittle: Synthesis and Electrochemical Performance of $La_{0.6}Ca_{0.4}Fe_{1-x}Ni_xO_3$ (x = 0.1, 0.2, 0.3) Material for Solid Oxide Fuel Cell Cathode.
J. Power Sources, 192, 63-69, 2009
– Authors: K. Vidal, L. M. Rodríguez-Martínez, L. Ortega-San-Martín, A. Martínez-Amesti, M. L. Nó, T. Rojo,
Laresgoiti, M. I. Arriortua.
Tittle: The Effect of Doping in the Electrochemical Performance of $(Ln_{1â``x}M_x)FeO_{3â``î´}$ SOFC Cathodes.
J. Power Sources,192, 175-179, 2009
– Authors: A. Martínez-Amesti, A. Larrañaga, L.M. Rodríguez, M.L. NÓ1, J.L. Pizarro, A. Laresgoiti, M.I. Arriortua.
Tittle: Chemical Compatibility Between YSZ and SDC Sintered at Different Atmospheres for SOFC Applications.
J. Power Sources, 192, 151-157, 2009
– Authors: I. de Meatza, I. Ruiz de Larramendi, J. I. Ruiz de Larramendi, M.I. Arriortua, T. Rojo.
Tittle: Nanostructure and Impedance Spectroscopy of $Pr_{0.7}Sr_{0.3}Mn_{1-x}M_xO_3$ (M= Fe, Co, Ni; x=0 and 0.2) thin Films Growth by Pulsed Laser Deposition
The Open Inorganic Chemistry Journal, 3, 47-55, 2009
– Authors: A. Martínez-Amesti, A. Larrañaga, L.M. Rodríguez-Martínez, M.L. Nó, J.L. Pizarro, A. Laresgoiti y
M.I. Arriortua.
Tittle: Influence of DSC-YSZ Contact at Different Atmospheres in SOFC Operation and Processing Conditions
Journal of the Electrochemical Society, 156 (7), B856- B 861, 2009
– Authors: J. Orive, E. Legarra, F. Plazaola, M. I. Arriortua, T. Rojo.
Tittle: Mild Hydrothermal Synthesis, Crystal Structure, Spectroscopic and Magnetic Properties of the $[M^{II}_xM^{III}_{2.5-x}(H_2O)_2(HP^{III}O_3)_y(P^VO_4)_{2-y}F]$ [M=Fe, x= 2.08, y= 1.58; M= Co, Ni, x= 2.5, y= 2]

J. Solid State Chem., 182, 2191-2201, 2009

- Authors: E. S. Larrea, J. L. Mesa, J. Rodríguez Fernández, M. I. Arriortua, T. Rojo.
  Tittle: Mild Hydrothermal Synthesis and Structural Determination of Two Layered, Strructuraly Related Inorganic Organic Hybrid Vanadates with Nickel (II) and Tris (2-Aminoetyl) amine.
  Eur. J. Inorg. Chem., 3607-3612, 2009
- Authors: R. FernÁndez de Luis, J.L. Mesa, M.K. Urtiaga, T. Rojo, M. I. Arriortua.
  TittleTwo Self-Catenated Nickel(II) Hybrid Vanadates with Honeycomb-Like 3D Inorganic Frameworks Stabilized by Crossed Organic Bpe Pillars: Thermal, Spectroscopic and Magnetic Properties.
  Eur. J. Inorg. Chem., 3607-3612, 2009
- Authors: A. Larrañaga, J.L. Mesa, J.L. Pizarro, M.I. Arriortua, T. Rojo.
  Tittle: Transition Metals Selenites: A Synthetic, Structural and Magnetic Revision.
  Book: Magnetic Materials: Research, Technology and Applications ISBN: 978-1-60692-145-6, Chapter. 5, 205-236, 2009
- Authors: A. Martínez-Amesti, A. Larrañaga, L. M. Rodriguez-Martinez, M. L. No, M. I. Arriortua.
  Tittle: Degradation of Symmetrical Cathode-Interlayer-Electrolyte Cells.
  ECS Transactions, 25 (2), 2867-2870, 2009
- Authors: N. Ortiz-Vitoriano, I. Ruiz de Larramendi, J. I. Ruiz de Larramendi, M. I.Arriortua, T. Rojo.
  Tittle: Effect of Electrolyte Contribution on the Electrochemical Behaviour of $Pr_{0.8}Sr_{0.2}Fe_{0.8}Ga_{0.2}O_{3}$.
  ECS Transactions, 25 (2), 2799-2806, 2009
- Authors: A. Ecija, K. Vidal, A. Larranaga, M.L. No, A. Laresgoiti, M.I. Arriortua.
  Tittle: Effect of the Syntehtic Method on the Structure and Electrical Conductivity of $La_{0.5}Ba_{0.5}FeO_{3}$ as SOFC Cathode Materials.
  ECS Transactions, 25 (2), 2639-2644, 2009
- Authors: K. Vidal, L.M. Rodriguez, L. Ortega, A. Martinez-Amesti, M.L. No, T. Rojo, A. Laresgoiti, M. I. Arriortua.
  Tittle: Performance of $(Ln_{0.5}M_{0.5})FeO_{3-ï}$• ¤ Perovskites as Cathode for SOFCs: Effect of Mean Radius of the A Site Cations.
  ECS Transactions, 25 (2), 2427-2434, 2009

  253.- Authors: T. Berrocal, J. L. Mesa, J. L. Pizarro, B. BazÁn, M. Iglesias, J. L. Vilas, T. Rojo, M. I. Arriortua.
  Tittle: Catalytic Performance of the High and Low Temperature Polymorphs of $(C_6N_2H_{16})_{0.5}[(VO)(HAsO_4)F]$: Structural, Thermal, Spectroscopic and Magnetic Studies.
  Dalton Trans., 39, 834-846, 2010
- Authors: A. Calderón, M.G. Barandika, B. Bazán, M.K. Urtiaga, M. I. Arriortua.
  Tittle: Unprecedented Coordination Modes for PDC (pyridine-2,5- dicarboxylate) in Polymorphic 3D Heterobimetallic Compounds alpha- and beta-$[MNa_2(PDC)_2(H_2O)_4]$, with M=Ni, Co.
  Cryst. Eng. Comm., 12(6),1784-1789, 2010
- Authors: R. Fernández de Luis, J.L. Mesa, M.K. Urtiaga, M. I. Arriortua, T. Rojo.

Tittle: Four Nodal Self Catenated $[\{Ni_8(Bpy)_{16}\}V_{24}O_{68}]\cdot 8.5(H_2O)$, Combining a Three Dimensional Metal-Organic and Inorganic Frameworks.
Cryst. Eng. Comm., 12(6), 1880-1886, 2010

- Authors: R. Fernández de Luis, M.K. Urtiaga, J.L. Mesa, M. I. Arriortua, T. Rojo.
  Tittle: Tris[4,4 '-(ethene-1,2-diyl)dipyridinium] Decavanadate Dihydrate
  Acta Cryst., E66, M323-U3773, 2010
- Authors: R. FernÁndez de Luis, M.K. Urtiaga, J.L. Mesa, K. Vidal, L. Lezama, T. Rojo, M. I. Arriortua.
  Tittle: Short-Range and Long-Range Magnetic Ordering, in Third Generation Brannerite Type Inorganic-Organic Vanadates: $[\{Mn(Bpy)\}(VO_3)_2]$â%o^$(H_2O)_{1.24}$ and $[\{Mn(Bpy)_{0.5}\}(VO_3)_2]$â%o^$(H_2\ O)_{0.62}$.
  Chem. Mater., 22(19), 5543- 5553, 2010
- Authors: K. Vidal, L.M. Rodriguez-Martinez, L. Ortega-San-Martin, M.L. Nó, T. Rojo, M. I. Arriortua
  Tittle: $(Ln_{0.5}M_{0.5})FeO_{3\text{-ï•}}$ ¤ Perovskites as Cathode for Solid Oxide Fuel Cells: Effect of Mean Radius of the A Site Cations.
  J. Electrochem. Soc., 157(8), 919- 924, 2010
- Authors: R. Cortés, N. de la Pinta, S. Martín, M.K. Urtiaga, G. Barandika, M. I. Arriortua. L. Lezama, G. Madariaga
  Tittle: Structural Analysis, Spectroscopic and Magnetic Properties of the 1D Triple-Bridged Compounds $[M(dca)_2(bpa)]$ (M= Mn, Fe, Co, Zn; dca= dicyanamide; bpa= 1,2-bis(4-pyridyl)ethane) and the 3D $[Ni(dca)(bpa)_2]dca.6H_2O$.
  Inorg. Chem., 49(22), 10445-10454, 2010
- Authors: I. Ruiz de Larramendi, R. Pinedo N. Ortiz-Vitoriano, J.I. Ruiz de Larramendi, M.I. Arriortua, T. Rojo
  Tittle: Nanostructured $Gd_{0.8}Sr_{0.2}Fe_{0.8}M_{0.2}O_3$ (M = Mr, Ga) Materials for Solid Oxide Fuel Cell Cathodes.
  Physics Procedia, 8, 2-9, 2010
- Authors: T. Berrocal, E. S. Larrea, M. Iglesias, M. I. Arriortua.
  Tittle: Vanadyl Arsenates as Catalysts for Selective Oxidation of Organic Sulfides and Alkenes.
  Molecular Catalysis A: Chemical, 335, 176-182, 2011

In: Chemistry Researcher Biographical Sketches
Editors: E. S. Huseynov and V. Babayev
ISBN: 978-1-62100-771-5

Chapter 7

BIOGRAPHICAL SKETCH

Name:	Title:	Date of Birth:
Bin Bao	Ph.D., M.D	

EDUCATION:

Institution and Location	Degree	Year Conferred	Scientific Field
Public Health, Tongji Medical University, Wuhan, China, July	M.D	1984	
Nutrition and Food Hygiene, Tongji Medical University, Wuhan, China, July	M.M.	1989	
Nutrition and Food Science, North Dakota State University, Fargo, ND, December	M.S	1992	
Nutrition and Food Science, Wayne State University, Detroit, MI, December	Ph.D	1997	

CONTACT POINTS:

Phone: 313-576-9796

E-mail: bbao@karmanos.org/ bbao@med.wayne.edu

Local: Excellent Student Awards, Tongji Medical University, annually 1980 to 1984
Rumble Scholarship, Wayne State University, Detroit, MI, 1993

National:Excellent Research Award, The Society of Maternal and Infant Nutrition, China, 1992
NIH Fellowship, Physician/Immunology Training Program, USA, 1999, 2001

Awards and Honors:

Local: Excellent Student Awards, Tongji Medical University, annually 1980 to 1984
Rumble Scholarship, Wayne State University, Detroit, MI, 1993

National: Excellent Research Award, The Society of Maternal and Infant Nutrition, China, 1992
NIH Fellowship, Physician/Immunology Training Program, USA, 1999, 2001

RESEARCH AND PROFESSIONAL EXPERIENCE:

Major Profeesional Societies

Chinese Society of Prevention, China
Chinese Society of Maternal Child Nutrition, China
Institution of Food and Technique, USA
American Society of Hematology, USA
American Association of Cancer Research, USA

Research Experience

07/87-03/91 - Researcher, Preventive Medicine and Pediatrics, Tongji Medical University, Wuhan, China
03/91-10/92 - Graduate Research Assistant, Food and Nutrition Department, College of Science, North Dakota State University, Fargo, ND
05/93-05/97 - Graduate Research Assistant, Department of Nutrition and Food Science, College of Science, Wayne State University, Detroit, MI
06/98-01/05 - Post-Doctoral Fellow/Clinic Investigator, Division of Hematology/Oncology, Department of Internal Medicine, Wayne State University School of Medicine, Detroit, MI
02/05- Research Associate/Research Assistant Professor, Division of Hematology/Oncology, Department of Internal Medicine, Karmanos Cancer Institute/Department of Pathology, Wayne State University School of Medicine, Detroit, MI

Grants

1. Co-Principal Investigator, "Effect of Homemade Formula on Growth and Development of Infants in
2. Rural Areas", World Health Organization, RMY$ 200,000 (US$ 20,000), 8/1988-3/1991
3. Co-Investigator (100%), "Zinc Deficiency and Th1 Functions: Molecular Mechanisms", NIH, $1,250,000, 07/2002-02/2007
4. PI, "Anti-angiogenic peptides for cancer treatment", DMCIOAD, $25,000, 02/01/2007-01/31/2008.
5. PI, "Synthetic anti-angiogenic peptides for cancer therapy", WSU, $10,000, 05/01/2007-04/30/2008.
6. Co-PI, "Zinc as an anti-inflammatory agent in diabetes mellitus", Harper University Hospital Medical Research Grant Program, $25,000, 07/1/2007-06/30/2008.
7. PI, " Regulation of HIF-1α, miR-200 and markers of cancer stem cells by CDF under hypoxic condition", Department of Defence (DOD), $75,000, 4/1/2011-4/30/2012 PC101482.
8. PI, " Trageting MnSOD through deregulation of miRNAs by novel agents", Department of Defence (DOD), $75,000, 4/1/2012-4/30/2013 (pending)

Publications

Peer-Reviewed Scientific Studies

1. Zhu, G.H., **Bao, B.** et al. (1987) The evaluation of nutritional value of instant noodles fortified with zinc, vitamins B_1, B_2, and C in school children. Food Sci 6:375-379.
2. **Bao, B.** and Zhao, W.Z. (1989) A relationship of exogenous obese children and their nutritional status. Acta Nutrimenta Sinica 9:221-227.
3. Zhao, W.Z., **Bao, B.** et al. (1989) The effect of high fiber food amorphophalus konjac genus on the serum lipid levels and body weight in rats with hypercholesterolemia induced by high-lipid diet. Acta Universitatis Medicinae Tongji 18:673-678.
4. **Bao, B.** and Chang, K.C. (1993) Carrot juice color, carotenoids, and nonstarchy polysaccarides as affected by processing conditions. J Food Sci 59:1155-1158.
5. **Bao, B.** and Chang, K.C. (1993) Carrot pulp chemical composition, color, and water-holding capacity as affected by blanching. J Food Sci 59:1159-1161.
6. Prasad, A.S., Fitzgerald, J., **Bao, B.,** Beck, F.W., and Chandrasekar, P.H. (2000) Treatment of common cold with zinc acetate lozenges. Ann Int Med 133:245-252
7. **Bao, B.,** Lee, D.Y., and Prasad, A. (2001) Isolation and characterization of mammary cDNA differentially expressed in lethal milk mutation mice. J Trace Elements in Exp Med. 14:371-382.
8. Prasad, A., **Bao, B.,** Beck, F.W. and Sarkar, F.H. (2001) Zinc activates NF-kB in HUT 78 cells. J Lab Clin Med. 138:250-256.
9. Prasad, A., **Bao, B.,** and Beck, F.W. (2002) Zinc enhances the expression of IL-2, IL-2 receptor alpha via NF-kB in HUT 78 cells. J Lab Clin Med 140:272-289.

10. **Bao, B**., Prasad, A., and Beck, F.W. (2003) Zinc modulates mRNA levels of cytokines. Am J Physiol. 285:E1095-E1102.
11. Prasad, A., **Bao, B.,** and Beck, F.W. (2004) Molecular evidence for anti-oxidant effect of zinc in human volunteers. Free Radical Biol Med, 37:1182-1190.
12. Prasad, A.S., **B Bao,** Beck, F.W., and Sarkar, F.H. (2006) Correction of IL-2 gene expression by in vitro zinc addition to MNC from zinc deficient human subjects: A specific test for zinc deficiency in humans. J. Lab. Clin. Med. 148:325-333.
13. Beck, F.W., Li, Y., **Bao, B.,** Prasad, A.S., Sarkar, F.H. (2006) Global gene expression in HUT-78 cells during zinc deficiency: Evidence for reprogramming the immune system. Nutrition, 22:1045-1056.
14. **Bao, B.,** Prasad, A., and Beck, F.W., Soneja, A., Sarkar, F.H. (2006) Toxic effect of zinc on NF-kB, IL-2, IL-2 receptor alpha, TNF-alpha in HUT-78 (Th0) cells. Toxic Lett. 166:222-228.
15. Prasad, A.S., Beck, F.W., **Bao, B.,** Fitzgerald, J.T., Snell, D.C., Steinberg, J.D., and Cardozo, L.J. (2007) Zinc supplementation decreases incidence of infections in the elderly: Effect of zinc on generation of cytokines and oxidative stress. Am J Clin Nutr. 85:837-844.
16. **Bao, B.,** Prasad, A.S., Beck, F.W. Suneja, A., and Sarkar, F. (2007) Molecular mechanisms: The role of zinc in NF-kB activation via IkB pathway in HUT-78 (Th_0) cells. FEBS Lett 581:4507-4511.
17. Prasad, A.S., Beck, F.W., **Bao, B.,** Fitzgerald, J.T., Snell, D.C. (2008) Duration and severity of symptoms and levels of plasma cytokines and adhesion molecules in patients with common cold treated with zinc acetate. J of Infectious Disease, 197:795-802.
18. **Bao, B.,** Prasad, A.S., Beck, F.W., Swerdlow, P., and Suneja, A. Sarkar, FH. (2008) Zinc supplementation decreases oxidative stress, incidence of infection, and generation of inflammatory cytokines in sickle cell disease patients. Translational Res 152:67-80.
19. **Bao, B.,** Prasad, A., Beck, F.W., Snell, D.C., Fitzgerald, J.T., Bao, G.W., Singh, T., Cardozo, L.J. (2010) Zinc decreases C-reactive protein, lipid peroxidation, and inflammatory cytokines in elderly subjects: a potential implication of zinc as an atheroprotective agent. Am J Clin Nutr 91: 1634-41.
20. Prasad, A.S., **Bao, B.,** Beck, F.W., and Sarkar, F.H. (2010) Zinc suppressed inflammatory cytokines by induction of A20-mediated inhibition of NF-κB. Nutrition (In press).
21. **Bao B**, Wang Z, Li Y, Kong D, Ali S, Banerjee S, Ahmad A, and Sarkar FH. (2010) The complexities of obesity and diabetes mellitus with the development and progression of pancreatic cancer. BBA Review on Cancers 1815:135-46.
22. Ali S, Ahmad A, **Bao B**, and Sarkar FH. (2011) Potential Molecular Link between Obesity, Diabetes and Cancer Obesity Review (Accepted).
23. Azmi, A., Wang, Z., **Bao, B.,** Banerjee, S., Ahmad, A., GunnJ.R., Yiwei Li, Y., Dejuan Kong, D., Ali,S., Sethi, S., Mohammad, R., Miele, L., Korc, M., Sarkar, FH, Ramzi Mohammad. (2011) Network Modeling and Validation of MDM2 inhibitor-Oxaliplatin Response Signatures in Solid Tumors. Science Translational Medicine. (Submitted).

24. Ahmad A, Z Wang, Wojewoda C, R Ali, D Kong, MY Maitah, S Banerjee, **Bao B,** S Padhye, and FH Sarkar (2011) Garcinol-induced apoptosis in prostate and pancreatic cancer cells is mediated by NF-kB signaling. Fronties in Bioscience. (In press).

25. **Bao B**, Shadan Ali S, Kong D, Sanila H. Sarkar, Zhiwei Wang, Sanjeev Banerjee, Subhash Padhye, Philip A. Philip and Fazlul H. Sarkar (2011) Anti-tumor activity of a novel compound-CDF is mediated by regulating miR-21, miR-200, PTEN in pancreatic cancer. PLOS One. 6(3):e17850

26. **Bao B,** Z Wang, S Ali, D Kong, Y Li, A Ahmad, S Banerjee, AS. Azmi, and FH. Sarkar (2011) Notch-1 induces Epithelial-mesenchymal transition consistent with cancer stem cell phenotype in pancreatic cancer cells. Cancer Lett (In press).

27. **Bao B,** Z Wang, S Ali, D Kong, S Banerjee, A Ahmad, Y Li, AS. Azmi, and FH Sarkar (2011) Over-expression of FoxM1 leads to epithelial-mesenchymal transition and cancer stem cell phenotype in pancreatic cancer cells. JCB (In press).

28. **Bao B,** Prasad AS, WJ Beck, Bao GW, Singh T, Ali S, Sarkar FH. (2011) Intracellular free zinc up-regulates IFN-γ and T-bet essential for Th_1 differentiation in Con-A stimulated HUT-78 cells. BBRC. 407:703-707.

29. Li Y, D Kong, Z Wang, S Ali, Y Li, A Ahmad, **Bao B,** and FH. Sarkar (2011) Inactivation of AR/TMPRSS2-ERG/Wnt signaling networks attenuates the aggressive behavior of prostate cancer cells. Cancer Prevention Research (Accepted).

30. Wang Z, Banerjee S, Ahmad A, Li Y, Azmi AS, Dr. Jason R Gunn , Kong D, **Bao B**, Ali S, Dr. Mohammad RM, Miele D, Korc M, Sarkar FH (2011) Activated K-ras and INK4a/Arf deficiency cooperate during the development of pancreatic cancer by activation of Notch and NF-kappaB signaling pathways. Plos One (Accepted).

31. Azmi A, Banerjee S, Ali S, Wang Z, **Bin Bao**, Frances WJ Beck, Anthony F Shields, Philip Philip, Fazlul H Sarkar, Ramzi M Mohammad (2011) Network Modeling of MDM2 Inhibitor-Oxaliplatin Combination Reveals Biological Synergy in wt-p53 solid tumors. Oncotargets (In press).

32. **Bao B,** S Ali, Z Wang, S Banerjee, A Ahmad, D Kong, Y Li, AS. Azmi, and FH Sarkar (2011) Metformin inhibits cell proliferation, migration and invasion by attenuating CSC function mediated by deregulating miRNAs in pancreatic cancer cells. Cancer Prevention Research (Submitted)

33. **Bao B,** S Ali, Z Wang, S Banerjee, A Ahmad, D Kong, Y Li, AS. Azmi, and FH Sarkar (2011) Cancer Research (submitted).

34. Li Y, D Kong, Z Wang, S, **Bao B**, A Ahmad, and FH. Sarkar (2011) Targeting the Hedgehog signalling pathway in cancer therapy. Expert Opinion On Therapeutic Targets (Submitted)

35. Shamim U, Hanif S, Albanyan A, Beck FW, Bao B, Wang Z, Banerjee S, Sarkar FH, Mohammad RM, Hadi SM, Azmi AS. (2011) Resveratrol-induced apoptosis is enhanced in low pH environments associated with cancer. J Cell Physiol (In press).

Chapter book

Bao B, A Ahmad, AS. Azmi, and FH Sarkar (2011) "The protective role of zinc in cancer: A potential chemopreventive agent" in *Nutrition, Diet and Cancer* (edited Dr. Rakesh Srivastava, First Edition), Springer Publisher (September, 2011)

Abstracts

1. Lee, D.Y. and Bao, B. (1993) Abnormal zinc metabolism in lactation lm mice. FASEB J 7:A773.
2. Lee, D.Y., Bao, B., and Klurfeld, D.M. (1994) Effect of dietary pectin and wheat bran on c-myc, c-jun-B and metallothionein messenger-RNA expression in colonic cells of rats. FASEB J 8:A492.
3. Lee, D.Y., Bao, B., and Klurfeld, D.M. (1995) Caloric restriction decreases the level of colonic thymidine kinase mRNA in rats. FASEB J 9:A990.
4. Bao, B. and Lee, D.Y. (1995) Use of mRNA differential display for characterization of murine lethal milk syndrome. FASEB J 9:A739.
5. Bao, B., Sun, Q., Klurfeld, D.M., and Lee, D.Y. (1996) Caloric restriction enhances the expression of hepatic metallothionein gene in rats with induced colon cancer. FASEB J. 10:A492.
6. Bao, B., Prasad, A., and Beck, F.W. (2001) Zinc enhances the expression of IL-2, IL-2 receptors in HUT 78 cells via NF-kB activation. Blood 98:32b
7. Bao, B, Prasad, A., and Beck, F. (2001) Zinc deficiency altered gene expression of IL-2, INF-gamma, TNF-alpha, and IL-8 in subgroup of lymphocytes. Blood 98:32b.
8. Bao, B., Prasad, A.S., Beck, F.W., Swerdlow, P., and Suneja, A. (2006) Beneficial effect of zinc supplementation on oxidative stress, cytokines, and NF-κB DNA binding in sickle cell disease patients. Blood 108:354a.
9. Bao, B., Prasad, A., Dou, D., Jin, S., Bao, G.W. (2007) Effect of synthetic peptides derived from human Apo(a) on cell proliferation, VEGF production, and angiogenesis in cancer cells and vascular endothelial cells. Blood 110:3904a
10. Bao, B., Prasad, A., Dou, D., Jin, S., Bao, G.W., Singh. (2008) Synthetic peptides derived from human Apo(a) inhibit cell proliferation, VEGF expression, NF-κB activation, and angiogenesis in cancer cells and vascular endothelial cells. Blood (online).

Presentations

1. Bao, B and Chang, S. (1992) Effect of blanching and retorting on carotenoids, non-starchy polysaccarides and color of carrot juice and pulp products. New Orleans, LA, Annual National Institute of Food Technology Conference.
2. Zhao, Y.P., Bao, B., and Chang, S. (1992) Effect of sulfite and starch treatments on quality of dehydrated carrots. New Orleans, LA, Annual National Institute of Food Technology Conference.

3. Lee, D.Y. and Bao, B. (1993) Abnormal zinc metabolism in lactation lethal milk mutation mice. Annual FASEB Conference.
4. Lee, D.Y., Bao, B., and Klurfeld, D.M. (1994) Effect of dietary pectin and wheat bran on c-myc, c-jun-B and metallothionein messenger-RNA expression in colonic cells of rats. Annual National FASEB Conference.
5. Lee, D.Y., Bao, B., and Klurfeld, D.M. (1995) Caloric restriction decreases the level of colonic thymidine kinase mRNA in rats. FASEB J 9:A990 (also presented)
6. Bao, B. and Lee, D.Y. (1995) Use of mRNA differential display for characterization of murine lethal milk syndrome. Annual National FASEB Conference.
7. Bao, B., Sun, Q., Klurfeld, D.M., and Lee, D.Y. (1996) Caloric restriction cnhanccs the expression of hepatic metallothionein gene in rats with induced colon cancer. Annual National FASEB Conference.
8. Bao, B., Prasad, A.S., Beck, F.W., Swerdlow, P., and Suneja, A. (2006) Beneficial effect of zinc supplementation on oxidative stress, cytokines, and NF-κB DNA binding in sickle cell disease patients. American Society of Hematology (ASH) Annual Meeting (December).
9. Bao B, Shadan Ali S, Kong D, Sanila H. Sarkar, Zhiwei Wang, Sanjeev Banerjee, Subhash Padhye, Philip A. Philip and Fazlul H. Sarkar (2011) Targeted killing of cancer stem-like cells by a novel analogue of curcumin. The 102nd AACR Annual Meeting, Orlando, FL
10. Wang Z, Banerjee S, Ahmad A, Li Y, Azmi AS, Dr. Jason R Gunn , Kong D, Bao B, Ali S, Dr. Mohammad RM, Miele D, Korc M, Sarkar FH (2011) Activated K-ras and INK4a/Arf deficiency cooperate during the development of pancreatic cancer by activation of Notch and NF-kappaB signaling pathways. The 102nd AACR Annual Meeting, Orlando, FL.
11. Bao B, S Ali, Z Wang, S Banerjee, A Ahmad, D Kong, Y Li, AS. Azmi, and FH Sarkar (2011) Metformin inhibits cell proliferation, migration and invasion by attenuating CSC function mediated by deregulating miRNAs in pancreatic cancer cells. American Pancreatic Cancer Association Annual Meeting, Chicago, IL

In: Chemistry Researcher Biographical Sketches
Editors: E. S. Huseynov and V. Babayev

ISBN: 978-1-62100-771-5

Chapter 8

BIOGRAPHICAL SKETCH

Name:	Title:	Date of Birth:
Bianco-Peled Havazelet	Department of Chemical Engineering, Technion - Israel Institute of Technology	07-10-1963

EDUCATION:

Institution and Location	Degree	Year Conferred	Scientific Field
D.Sc. Chemical Engineering			

CONTACT POINTS:

Technion City, Haifa 32000 Israel

RESEARCH AND PROFESSIONAL EXPERIENCE:

Biomaterials, Drug delivery, tissue engineering

Professional Appointments:

Associate Professor

Honors:
1986 - Chemical Engineering Department, Distinction Award.
1987, 1988 - Technion Undergraduate School, Distinction Award.

1995 -Technion, Distinction Award for Teaching Assistants.
1997 - Gutwirth Fellowship, Special Distinction Award
1997 - Fulbright Foundation, Post-Doctoral Fellowship.
2000, 2001 - Torchinsky Family Academic Fund - Canada, Lectureship.
2001 - Henri Gutwirth Fund for the Promotion of Research, Research Award
2007 - Hershel Rich Technion Innovation Award
2007 - Nomination as an AMIT (Alfred Mann Institute) Faculty Member
2010 - Daniel Shiran Memorial Research Award

Publications during last three years

1. S. Zabar, E. Shimoni and H. Bianco-Peled, "Development of Nanostructure in Resistant Starch Type III during Thermal Treatments and Cycling", *Macromolecular Bioscience* 8(2), 163-170, 2008.
2. D. Dikovsky, H. Bianco-Peled and D. Seliktar, "Defining the Role of Matrix Compliance and Proteolysis in Three-Dimensional Cell Spreading and Remodeling", *Biophysical Journal* 94(7), 2914-2925, 2008.
3. R. Bitton and H. Bianco-Peled, "Novel Biomimetic Adhesives Based on Algae Glue", *Macromolecular Bioscience* 8(5), 393-400, 2008.
4. J. Hedlund, M. Andersson, C. Fant, R. Bitton, H. Bianco-Peled, H. Elwing and M. Berglin, "Change of Colloidal and Surface Properties of Mytilus edulis Foot Protein 1 in the Presence of an Oxidation ($NaIO_4$) or a Complex-Binding (Cu^{2+})", *Biomacromolecules* 10(4), 845-849, 2009.
5. R. Bitton, E. Josef, I. Shimshelashvili, K. Shapira, D. Seliktar and H. Bianco-Peled, "Phloroglucinol-based biomimetic adhesives for medical applications", *Acta Biomaterialia* 5(5), 1582-1587, 2009.
6. M. Davidovich-Pinhas, O. Harari and H. Bianco-Peled, "Evaluating the mucoadhesive properties of drug delivery systems based on hydrated thiolated alginate", *Journal of Controlled Releases* 136(1), 38-44, 2009.
7. S. Zabar, U. Lesmes, I. Katz, E. Shimoni and H. Bianco-Peled, "Studying Different Dimensions of Amylose-Long Chain Fatty Acid Complexes: Molecular, Nano and Micro Level Characteristics", *Food Hydrocolloids* 23(7), 1918-1925, 2009.
8. I. Frisman, R. Orbach, D Seliktar and Havazelet Bianco-Peled, "Structural investigation of PEG-fibrinogen conjugates", *Journal of Materials Science: Materials in Medicine* 21(1), 73-80, 2010.
9. M. Davidovich-Pinhas and H. Bianco-Peled, "A quantitative analysis of alginate swelling", *Carbohydrate Polymers* 79(4), 1020-1027, 2010.
10. M. Davidovich-Pinhas and H. Bianco-Peled, "Mucoadhesion: a review of characterization techniques", *Expert Opinion on Drug Delivery* 7(2), 259-271, 2010. (Invited paper).
11. S. Zabar, U. Lesmes, I. Katz, E. Shimoni and H. Bianco-Peled, "Structural characterization of amylose-long chain fatty acid complexes produced via the acidification method ", *Food Hydrocolloids* 24(4), 347-357, 2010.

12. O. Yom Tov, S. Luvitch and H. Bianco-Peled, "Molecularly imprinted hydrogel displaying reduced non-specific binding and improved protein recognition", *Journal of Separation Science* 33(8), 1-9, 2010.
13. Frisman, D Seliktar and Havazelet Bianco-Peled, "Nanostructuring of PEG-Fibrinogen Polymeric Scaffold", *Acta Biomaterialia* 6(7), 2518-2524, 2010.
14. D. Dikovsky, H. Bianco-Peled and D. Seliktar, "Proteolytically Degradable Photo-Polymerized Hydrogels Made from PEG-Fibrinogen Adducts", *Advanced Enginnering Materials* 12(6), B200-B209, 2010.
15. M. Davidovich-Pinhas and H. Bianco-Peled, "Novel mucoadhesive system based on sulfhydryl-acrylate interactions", *Journal of Materials Science: Materials in Medicine* 21(7), 2027-2034, 2010.
16. E. Josef, M. Zilberman and H. Bianco-Peled, "Composite Alginate Hydrogels: An Innovative Approach for Controlled Release of Hydrophobic Drugs", *Acta Biomaterialia* 6(12), 4642-4649, 2010.
17. R. S. Tu, R. Marullo, R. Pynn, R. Bitton, H. Bianco-Peled and M. V. Tirrell, “Cooperative DNA binding and assembly by a bZip peptide-amphiphile”, *Sotf Matter* 6(5), 1035-1044, 2010.
18. M. Davidovich-Pinhas and H. Bianco-Peled, "Alginate-PEGAc: A new mucoadhesive polymer", *Acta Biomaterialia* 7(2), 625-633, 2011.
19. M. Davidovich-Pinhas and H. Bianco-Peled, “Physical and structural characteristics of acrylated poly(ethylene glycol)-alginate conjugates”, *Acta Biomaterialia* 7(7), 2817-2825, 2011.
20. Frisman, J. Shachaf, D. Seliktar and H. Bianco-Peled, "Stimulus-Responsive Hydrogels Made From Biosynthetic Fibrinogen Conjugates for Tissue Engineering: Structural Characterization", *Langmuir,* in press.
21. Frisman, D. Seliktar and H. Bianco-Peled, “Nanostructuring PEG-Fibrinogen Hydrogels to Control Cellular Morphogenesis”, *Biomaterials*, in press.

In: Chemistry Researcher Biographical Sketches
Editors: E. S. Huseynov and V. Babayev
ISBN: 978-1-62100-771-5

Chapter 9

BIOGRAPHICAL SKETCH

Name:	Title:	Date of Birth:
Vladimir Burtman	Geophysics department, University of Utah, in Salt Lake City, USA	Nov-02-1965

EDUCATION:

Institution and Location	Degree	Year Conferred	Scientific Field
School of Applied Science at Hebrew University of Jerusalem, Israel	Ph.D.	(1996)	

CONTACT POINTS:

Frederick Albert Sutton Building 115 S 1460 East Room 383 Salt Lake City, UT 84112

Professional Appointments:

Research Professor of Physics, Project Director in Consortium for EM Inversion (CEMI) studies, both in University of Utah

Awards and Honors:

2011-2013 - $1.2 million award of AMIRA International research grant: Spectral Induced Polarization for 3D Mineral Discrimination (P1058, Co-PI);
2007 - Research Council Award, Physics Department, University of Utah,

2006 - Award of research Nano-engineering research (NER) NSF grant: Spintronics in self-assembled monolayers (Co-PI);

1994 - Sally Berg Postdoctoral Fellowship, The Hebrew University of Jerusalem, Jerusalem, Israel.1996; Fellowship Ben- Gurion fund, Ph D, Ministry of science & technology, Jerusalem, Israel.

RESEARCH AND PROFESSIONAL EXPERIENCE:

Applied Solid State Physics, Molecular Electronics, Geophysics

PUBLICATIONS LAST 3 YEARS:

(A) Physics:

V. Burtman, 2011, Molecular Orbital Gap Studies in Tunneling Single Molecular Devices, Nova Scientific Publisher House, Series: Physics Research and Technology, ISBN: 978-1-61122-786-4.

V. Burtman, A. Zelichonok and A. V. Pakoulev, 2011, Molecular Photovoltaics in Nanoscale Dimension (Review): International Journal of Molecular Sciences, 12, 173-225; doi:10.3390/ijms12010173.

V. Burtman, 2011, Transport Studies in Molecular Devices: Chapter 6 in Molecular Self-Assembly: Advances in Chemistry, Biology and Nanotechnology, Editors: James P. Comrie, Series: Chemical Engineering Methods and Technology; Biotechnology in Agriculture, Industry and Medicine, ISBN: 978-1-61122-412-2

A. V. Pakoulev, V. Burtman, 2010, Study of the transport mechanism in molecular self-assembling devices (Invited paper): Applied Physics A, Volume 98, Issue 4, pp.717-734, DOI 10.1007/s00339-009-5531-5.

A. V. Pakoulev, D. Zaslavsky, V. Burtman, 2010, Photocurrent Study of the Transport Mechanism in Molecular Self-Assembling Field Effect Transistors: Transistors: Chapter 6 in Types, Materials and Applications, pp.133-174, Series: Electrical Engineering Developments, Editor: Benjamin M. Fitzgerald, ISBN: 978-1-61668-908-7.

A. V. Pakoulev, D. Zaslavsky and V. Burtman, 2009 Photocurrent Study of the Transport Mechanism in Molecular Self-Assembling Field Effect Transistors; Chapter 5 in Integrated Circuits, Photodiodes and Organic Field Effect Transistors, .pp. 115-159, Editors: Robert McIntire and Pierre Donnell, ISBN: 978-1-60692-660-4.

A. V. Pakoulev, V. Burtman, 2009, Temperature Dependent Barrier Crossover Regime in Tunneling Single Molecular Devices Based on the Matrix of Isolated Molecules: Journal of Physical Chemistry C, 113 (51), pp 21413–21421, DOI: 10.1021/jp9056576.

(B) Geophysics:

V. Burtman, M. Endo, M. S. Zhdanov, Thomas Ingeman-Nielsen, 2011, High-frequency induced polarization measurements of hydrocarbon-bearing rocks: SEG Expanded Abstracts 30, 4pages.

V. Burtman, M. Endo, M. S. Zhdanov, Thomas Ingeman-Nielsen, 2011, Induced polarization measurements of hydrocarbon-bearing rocks: : Consortium for Electromagnetic Modeling and Inversion, Proceedings of 2011 Annual Meeting, Salt Lake City, UT, 3-27.

V. Burtman, M. Endo, L. Fu, and M. S. Zhdanov , 2011. Generalized effective-medium theory of time-domain induced polarization: Theory, modeling, and applications: Consortium for Electromagnetic Modeling and Inversion, Proceedings of 2011 Annual Meeting, Salt Lake City, UT, 27-52.

V. Burtman, A. V. Gribenko, and M. S. Zhdanov , 2010, Advances in experimental research of induced polarization effect in reservoir rocks: SEG Expanded Abstracts 29, 2475-2479, doi:10.1190/1.3513351.

Burtman, V., A. Gribenko, and M. S. Zhdanov, 2010, Advances in experimental research of induced polarization effect in reservoir rocks: Proceedings of the CEMI 2010 Annual meeting, 53-72.

V. Burtman, A. V. Gribenko, and M. S. Zhdanov , 2009, Induced polarization in hydrocarbon-saturated sands and sandstones: experimental study and general effective medium modeling: SEG Expanded Abstracts 28, 774-778.

M. S. Zhdanov, A. Gribenko, V. Burtman and V. I. Dmitriev,2008, Anisotropy of induced polarization in the context of the generalized effective-medium theory: SEG Expanded Abstracts 27, 677-682; doi:10.1190/1.3063740.

In: Chemistry Researcher Biographical Sketches
Editors: E. S. Huseynov and V. Babayev
ISBN: 978-1-62100-771-5

Chapter 10

BIOGRAPHICAL SKETCH

Name:	Title:	Date of Birth:
Vito Capozzi	Department of Biomedical Sciences, University of Foggia	27-04-1950

EDUCATION:

Institution and Location	Degree	Year Conferred	Scientific Field
	Master degree in Physics and Ph. D. in Physics		

CONTACT POINTS:

Viale L. Pinto, I-71100 Foggia (Italy)

RESEARCH AND PROFESSIONAL EXPERIENCE:

Raman spectroscopy of layered semiconductors.

Optical properties and photoluminescence of localized excitons in III-VI semiconductors. Stimulated emission and optical gain measurements in layered semiconductors.

Optical and morphological properties of nanoaggregates of solid particles of a-C, graphite and a-Si. Optical and structural properties of film and crystal Fullerenes. Study of Anderson and Stark localization in disordered GaAs/AlGaAs superlattices.

Optical, structural and electrical properties of II-VI semiconductors deposited by "laser ablation" technique.

Recently, he is interested to the chemical-physics, optical and structural properties of biological molecules and biopoliymers having large interest in biomedical applications.

Professional Appointments:

Vito Capozzi was graduated in Physics in 1974 at the University of Bari. From 1974 to 1978 he had a research contract at the "Libera Università di Trento".

He obtained the PhD degree in Physics in 1978.

From September 1978 to march 1982 he was research assistant at the Institute of Applied Physics, Federal Institute of Technology in Lausanne (CH).

In 1982 he had a researcher position in "Condensed Matter Physics" at the University of Trento. From 1987 to 1998 he was University senior researcher in "Condensed Matter Physics" at the Physics Department, University of Bari.

From 1998 to 2001 he was Associated Professor of Physics at the Medicine Faculty, University of Foggia, where, he is full professor of Applied Physics since 2001.

Publications during last three years

1. V. Capozzi, G. Mariotto, M. Montagna, A. Cingolani, A. Minafra: "Space-Charge-Limited Currents in GaSe at Different Temperatures" Phys. Status Solidi(a) 40, (1976) 93-100.
2. V. Capozzi, A.Fontana, MP.Fontana, G.Mariotto, M.Montagna, G.Viliani "Raman Scattering in PbI_2" Il Nuovo Cimento 39B, (1977) 556-560.
3. M.P. Fontana, G. Viliani, M. Montagna, V. Capozzi, G. Mariotto: "Raman Spectroscopy of TlCl in the Melting Point Region"Phys. Lett. 64A, (1977) 319-321.
4. A.Fontana, G.Mariotto, M.Montagna, V.Capozzi, E.Cazzanelli, M.P.Fontana:"Temperature Dependence of Anomalous and Ordinary Raman Scattering in β-AgI"Sol. State Commun. 28, (1978) 35-37.
5. L. Baldassarre, V. Capozzi, G. Maggipinto, A. Minafra:"Space-Charge-Limited Current Measurements in $ZnIn_2S_4$"Phys. Status Solidi(a) 46, (1978) 589-593.
6. Gobby, JL. Staehli, M. Guzzi, V. Capozzi:"Spectres Exitoniques et Propriet. Structurales de GaSe"Helv. Physica Acta 52, (1979) 337-341.
7. V. Capozzi, M. Di Giulio, S. Galassini, C. Manfredotti, A. Tepore:"Lifetime and Diffusion Length of Electrons in p-GaSe by Photovoltaic Effect Measurements at Room Temperature"Physica 105B, (1981) 50-53.
8. V.Capozzi: "Direct and Indirect Excitonic Emission in GaSe"Phys. Rev. B23, (1981) 836-840.
9. V. Capozzi, and A. Minafra:"Photoluminescence Properties of Cu-doped GaSe"J. Phys. C: Solid State Phys. 14, (1981) 4335-4346
10. V. Capozzi and A. Minafra:"Photoluminescence of GaSe:Cu Doped by Ion Implantation"Sol. State Commun. 38, (1981) 341-343.
11. V.Capozzi JL.Staehli, M.Guzzi:"A Simple Apparatus for the Direct Measurement of Optical Gain Spectra"Optics Commun. 38, (1981) 284-288; also mentioned inLaser Focus 17, (1981) 105-106.
12. JL. Staehli, V. Capozzi, M. Guzzi:"Gain Spectra of Gallium Selenide"Physica 105B, (1981) 35-39.

13. V. Capozzi and JL. Staehli:"Spontaneous and Optically Amplified Luminescence from Exciton-Exciton Collision in GaSe at Liquid-He Temperature"Phys. Rev. B28, (1983) 4461-4467.
14. V. Capozzi:"Kinetics of Radiative Recombinations in GaSe andInfluence of Cu-doping on the Luminescence Spectra"Phys. Rev. B28, (1983) 4620-4627.
15. V. Capozzi, S. Caneppele, M. Montagna:"Luminescence of Selective Optical Pumping in GaSe"J. of Luminescence 31, (1984) 463-465
16. V. Capozzi:"Luminescence Investigation of Direct and Indirect Excitons Bound to Deep-Neutral Acceptors in in ε-GaSe"Helv. Phys. Acta 58, (1985) 281-292. (invited papcr)
17. JL. Staehli and V. Capozzi:"Simultaneous Direct and Indirect Recombinations in the Electron-Hole Plasma of GaSe"Helv. Phys. Acta 58, (1985) 262-271.
18. 18-V.Capozzi, S. Caneppele, M.Montagna and F.Lèvy:"Radiative Decay from Free and Bound Excitons in GaSe"Phys. Status Solidi(b) 129, (1985) 247-257.
19. V. Capozzi, and M. Montagna:"Dynamics of Excitons in GaSe as a Function of Temperature"J. de Physique 46, (1985) 191-195.
20. V. Capozzi and K. Maschke:"Thermalization of Photoexcited Localized Excitons in GaSe Samples with Stacking Disorder": Phys. Rev. B34, (1986) 3924-3931.
21. L. Colangeli, V. Capozzi, E. Bussoletti, A, Minafra:"Extinction Spectra of Amorphous Carbon Submicronic Grains" Astronomy and Astrophysics 168, (1986) 349-353.
22. L. Pavesi and V. Capozzi:"Evidence of the Exciton-Plasma Transition in Emission Spectr of GaSe"Il Nuovo Cimento 8D, (1986) 531-540.
23. L. Pavesi, V. Capozzi, JL. Staehli:"Analysis of the Excitonic Mott Transition in Gallium Selenide"Sol. State Communic. 61, (1987) 321-325.
24. L. Pavesi, JL. Staehli, V. Capozzi:"On the Excitonic Mott Transition in the Static Screening Approximation"J. Phys. C: Sol. State Physics 21, (1988) 1485-1495.
25. L. Pavesi, V. Capozzi, JL. Staehli:"Red-shift of the Excitonic Level in Two Beam Absorption Spectra of GaSe"Physica Scripta 38, (1988) 627-631.
26. V. Capozzi, A. Minafra, L. Colangeli, E. Bussoletti, A. Borghesi:"Amorphous Carbon Grains: Size Effects in Ultraviolet Spectra"Sol. State Commun. 65, (1988) 1193-1196.
27. V. Capozzi, C. Flesia, A. Minafra:"Anomalous Dips in Extinction Spectra of Disordered Carbon Grains"Optics Communic. 68, (1988) 166-1170.
28. V. Capozzi, C. Flesia, A. Minafra:"Strong Localiz. of Light in Spongy Agglomerates of Carbon Grains"Physica Scripta 39, (1988) 63-66.
29. L. Pavesi, JL. Staehli, V. Capozzi:"Mott Transition of the Excitons in GaSe"Phys. Rev. B39, (1989) 10982-10994.
30. V. Capozzi and M. Montagna:"Optical Spectroscopy of Extrinsic Recombin. in Gallium Selenide"Phys. Rev. B40, (1989) 3182-3190.
31. G.F. Lorusso, A. Minafra, V. Capozzi, D. Lojacono:"Correlation Spectroscopy of Solid Microparticles Suspended in Aquous Medium".Helv. Physica Acta 62, (1989) 900-902.
32. V. Capozzi, L. Pavesi, JL. Staehli:"Luminescence from Exciton-Frre Carrier Collisions in GaSe".Journal of Luminescence 48, (1990) 111-115.
33. Y. Burki, P. Schwendimann, V. Capozzi W. Czaja:"Optical Gain in ZnTe:O at Room Temperature"Europhysics Letters 16, (1991) 763-766.
34. G.F. Lorusso, V. Capozzi, A. Minafra:"A Low Cost Correlator Employing a Personal Computer"Rev. of Scientific Instrum. 63, (1992) 2152-2156.

35. G.F. Lorusso, V. Capozzi, A. Minafra:"Experimental Analysis of the noise in photon correlationand photon structure functions"Phys. Rev. A45, (1992) 3391-3393.
36. G.F. Lorusso, A. Minafra, V. Capozzi:"Noise Dependence of Correlation and Structure Functions on Mean Count Rate and Sampling Time"Applied Optics 32, (1993) 3867-3870.
37. V. Capozzi, L. Pavesi, JL. Staehli:"Exciton-Carrier Scattering in Gallium Selenide"Phys. Rev. B47, (1993) 6340-6349.
38. G.F. Lorusso,V. Capozzi, P. Milani, A. Minafra, D. Lojacono:"UV Spectra of Graphite Microparticles Produced by Laser vaporization"Solid State Commun. 85 (1993) 729-734.
39. G.F. Lorusso, V. Capozzi, V. Augelli, A. Minafra, T. Ligonzo,C. Flesia, G. Bruno, P. Capezzuto:"Interference Effects in UV-Extinction Spectra of Inhomogeneous Amorphous Silicon".Phys. Rev. B48, (1993) 12292-12295.
40. G.F. Lorusso, A. Minafra, V. Capozzi:"Experimental Photon Statistics of Multiscattered Light"Phys. Rev. E49, (1994) 3531-3534.
41. Y. Burki, W. Czaja, V. Capozzi, P. Schwendimann:"Temperature Dependence of the Photoluminescence and lifetime of ZnTe:O"J. Phys.: Condens. Matter 5, (1993) 9235-9252.
42. Y. Burki, W. Czaja, V. Capozzi, P. Schwendimann:"Dynamics of Multiphonon Processes in Luminescence Spectra of Oxygen Bound Excitons in ZnTe:O"J. Luminescence 58, (1994) 206-209.
43. Y. Burki, W. Czaja, V. Capozzi, P. Schwendimann:"Fluorescent Decay Times of Excitons Bound to Oxygen Traps in ZnTe:O"J. Luminescence 60, (1994) 4-7.
44. G. Guizzetti, F. Marabelli, M. Patrini, V. Capozzi, G.F. Lorusso, A. Minafra, M. Manfredini, P. Milani:"Optical Properties of Fullerite Thin Films in the 0.4-32 eV Energy Range"Phys. Status Solidi (b) 183, (1994) 267-275,.
45. T. Ligonzo, V. Augelli, A. Minafra, V. Capozzi, D. Martin,G.F. Lorusso, P. Favia:"Photocurrent Spectroscopy in n-i-n GaAs/AlGaAs Short Period Superlattices".Solid State Commun. 94, (1995) 429-433.
46. G.F. Lorusso, V. Capozzi, J.L. Staehli, C. Flesia, D. Martin, P. Favia:"Absorption Spectra of GaAs/AlGaAs Random Superlattices at 2 K"Phys Rev B53, (1996) 1018-1021.
47. G.F. Lorusso, V. Capozzi, J.L. Staehli, C. Flesia, D. Martin, P. Favia, G. Perna:"The Role of Randomly Distributed Well Withs in Disordered GaAs/AlGaAs Superlattices"Semicond. Science & Tecnol. 11, (1996) 308-314.
48. G.F. Lorusso, F. Tassone, V. Capozzi, G. Perna, D. Martin:"Reflectance of GaAs/AlGaAs Disordered Superlattices"Solid State Commun. 98, (1996) 705-709.
49. V. Capozzi, G.F. Lorusso, J.L. Staehli D. Martin, G. Perna:"Temperature and Excitation Intensity Dependencies of the PL Spectra of GaAs/AlGaAs Disordered Superlattices"Phys. Rev. B54, (1996) 7643-7646.
50. G.F. Lorusso, F. Tassone, V. Capozzi, J.L. Staehli, D. Martin:"Anderson and Stark Localization in Superlattices: Residual Coupling and Fine Structure"Solid State Commun. 100, (1996) 17-20.
51. V. Capozzi, G. Casamassima, G.F. Lorusso, A. Minafra, R. Piccolo, T. Trovato, A Valentini:"Optical Characterization of Fullerite C60 Thin Films"Synthetic Metals 77, (1996) 3-5.

52. V. Capozzi, G. Casamassima, G.F. Lorusso, A. Minafra, R. Piccolo, T. Trovato, A Valentini:"Optical Spectra and Photoluminescence of C60 Thin Films"Solid State Commun. 98, (1996) 853-858.
53. G. Bruno, M. Losurdo, P. Capezzuto, V. Capozzi, G.F. Lorusso, A. Minafra:"Epitaxial Growth and processing of InP Films in a Novel Remote Plasma-MOCVD Apparatus"Materials Science Forum 203, (1996) 85-90.
54. G. Bruno, M. Losurdo, P. Capezzuto, V. Capozzi, T. Trovato, G.Perna G.F. Lorusso:"Hydrogen Plasma Passivation of InP: Real Time Ellipsometry Monitoring and ex-situ Photoluminescence Measurements"Appl. Phys. Lett. 69, (1996) 685-687.
55. G. Cicala, P. Capezzuto, G. Bruno, L. Schiavulli, V. Capozzi:"Plasma Dcposition and Characterization of Photoluminescent Fluorinated Nanocrystalline Silicon Films"J. Appl. Phys. 80, (1996) 6564-6566.
56. Giardini, M. Ambrico, D. Smaldone, R. Martino, V. Capozzi, G. Perna:"Laser Ablation of Highly Oriented CdSe Thin Films and CdSe/CdTe Multilayers on Silicon Substrates"Applied Surface Science 106, (1996) 144-148.
57. Giardini, M. Ambrico, D. Smaldone, R. Martino, V. Capozzi, G. Perna, G.F. Lorusso:"Structural and Optical Properties of II-VI Compounds Thin Films and II-VI Multilayer Structures grown on Si by Laser Ablation"Mater. Science & Eng. B43, (1997) 102-107.
58. V. Capozzi, T. Trovato, G. Casamassima, H. Berger, G.F. Lorusso:"Photoluminescence Spectra of C60 Thin Films Deposited on Different Substrates"Carbon 35, (1997) 763-766.
59. V. Capozzi, G.F. Lorusso, T. Trovato, B. Viguier, A. Minafra:"Influence of the Substrate Temperature on the Photoluminescence of C60 Films" J. Luminescence 74, (1997) 473-475.
60. V. Capozzi, G.F. Lorusso, G. Perna, A. Bitz:"Photoluminescence of Random GaAs/AlGaAs Superlattices". J. Luminescence 72, (1997) 361-363.
61. G.F. Lorusso, V. Capozzi, J.L. Staehli:"Extended States in One-Dimensional Random Potential"Solid State Comm. 103, (1997) 15-18.
62. V. Capozzi, T. Trovato, G. Perna, M. Losurdo, P. Capezzuto, G Bruno"Enhanced Photoluminescence from InP Deposited and Treated by Remote Plasma Processes"J. Luminescence 72, (1997) 98-100.
63. G. Perna, M. Ambrico, V. Capozzi, D. Smaldone, R. Martino:"Photoluminescence Analysis of Laser deposited CdS Films" J. Luminescence 72, (1997) 90-92.
64. M. Mobarak, H. Berger, G.F. Lorusso, V. Capozzi, G. Perna, MM. Ibrahim, G. Margaritondo:"The Growth and characterization of GaInSe2 single Crystals"Phys. D: Appl. Phys. 30, (1997) 2509-2513.
65. M. Mobarak, H. Berger, G.F. Lorusso, V. Capozzi, G. Perna, MM. Ibrahim, G. Margaritondo:"The Growth and Properties of Single Crystals of GaInTe2, a Ternary Calcogenide Semiconductor"Phys. D: Appl. Phys. 31, (1998) 1433-1437.
66. G.F. Lorusso, F. Tassone, V. Capozzi, D. Martin, J.L. Staehli:"Anderson and Stark Localization in GaAs/AlGaAs Disordered SL"Superlattices and Microstructures 23, (1998) 9-12.
67. G.F. Lorusso, V. Capozzi, JL. Staehli:"Photoluminescence Excitation Spectra of Disordered Superlattices"J. Luminescence 78, (1998) 259-264.

68. G. Perna, V. Capozzi, M. Ambrico:"Structural Properties and Photoluminescence Study of CdSe/Si Epilayers Deposited by Laser Ablation" J. Appl. Phys. 83, (1998) 3337-3344.
69. G. Perna, V. Capozzi, M. Ambrico, D. Smaldone:"Temperature Dependence of the Red-shift and Broadening of the Exciton Line in CdSe/GaAs Laser Ablated Heterostructures" J. Luminescence 77, (1998) 534-539.
70. M. Ambrico, D. Smaldone, V. Stagno, G. Perna, V. Capozzi: "Structural and Optical Parameters of ZnS_xSe_{1-x} Films Deposited on Quartz Substrates by Laser Ablation" Semicond. Sci. & Technol. 13, (1998) 1446-1455.
71. V. Capozzi, M. Santoro, G. Celentano, H Berger, A. Minafra: "Growth and Photoluminescence Spectra of C60 Single Crystals" J. Luminescence 76, (1998) 395-398.
72. G.F. Lorusso, V. Capozzi, F. Tassone, P. Favia, JL. Staehli: "Low Temperature Photocurrent Spectra of Ordered and Disorder Superlattices" Solid State Commun. 109, (1999) 305-310.
73. G. Perna, S. Pagliara, V. Capozzi, M. Ambrico, T. Ligonzo:"Optical Characterization of CdS_xSe_{1-x} Films Grown on Quartz Substrate by Pulsed Laser Ablation Technique"Thin Solid Films 349, (1999) 220-224.
74. V. Capozzi, G. Perna, S. Pagliara, M. Ambrico: "Absorption and Photoluminescence Analysis of CdS Ablated Films Grown on Quartz Substrates" Applied Surface Science 154, (2000) 283-243.
75. V. Capozzi, G. Celentano, G. Perna, G.F. Lorusso, A. Minafra:"Photoluminescence Properties of C60 Films Deposited on Si Substrate" J. Luminescence 86, (2000) 129-135.
76. G. Perna, S. Pagliara, V. Capozzi, M. Ambrico, M. Pallara:"Excitonic Luminescence of CdS_xSe_{1-x} Films Deposited by Laser Ablation on Si Substrate" Solid State Communications 114, (2000) 161-166.
77. G. Perna, V. Capozzi, G.F. Lorusso, G. Bruno, M. Losurdo, P. Capezzuto:"Photoluminescence Properties of Homoepitaxial InP Films Grown by Remote Plasma MOCVD Technique" Semicond. Science & Technol. 15, (2000) 736-743.
78. Scanni, A. Valentini, G. Perna, V. Capozzi, A. Convertino:"Photoluminescence Analysis on Teflon Bulk and Teflon-like Films Grown by Ion-beam Sputtering" J. Luminescence 91, (2000) 87-89.
79. S. Pagliara, L. Sangaletti, L.E. Depero, V. Capozzi, G. Perna:"Effect of Disorder on the Raman Scattering of CdSxSe1-x FilmsDeposited by Laser Ablation" Solid State Communications 116, (2000) 115-119.
80. V. Marotta, S. Orlando, A. Giardini, V. Capozzi, G. Perna, A.M. Santoro: "Electrical and Optical Characterization of Multilayered Thin Film Based on Pulsed Laser Deposition of Metal Oxides" Appl. Surf. Science 168, (2000) 141-145.
81. R. De Leo, V. Capozzi, C. Casalino, E. Cisbani, C. Coluzza, S. Frullani, F. Garibaldi, M. Iodice, L. Lagamba, E. Nappi, G.Perna, G.M. Urciuoli:"Chromatic Aberration and Forward scattering of Light in Aerogel"Nuclear Instruments & Methods A 457, (2001) 52-63.

82. G Perna, V. Capozzi, S. Pagliara, M. Ambrico, D. Lojacono: "Reflectance and Photoluminescence Characterization of CdS and CdSe Heteroepitaxial Film Deposited by Laser Ablation Technique" Thin Solid Films, 387, (2001) 208-211.
83. V. Capozzi, A.M. Santoro, G. Perna, A. Minafra, G. Casamassima: "Ecitonic Photoluminescence Spectra of C60 Single Crystals Grown by different Techniques" European Physics Journal: Applied Physics 14, (2001) 3-11.
84. G. Perna, V. Capozzi, V. Augelli, T. Ligonzo, L. Schiavulli, G. Bruno, M.Losurdo, P. Capezzuto, J.L. Staehli, M. Pallara:"Luminescence Study of the Disorder in Polycrystalline InP Thin Films"Semiconductor Science & Technology, 16, (2001) 377-385.
85. S.P. Kingsley, P.F. Biagi, R. Piccolo, V. Capozzi, A. Ermini, Y.M. Khatkevich: "Hydrogeochemical Precursors of Strong Earthquakes:A Realistic Possibility in Kamchatka" Phys. Chem. Earth (C) 26, (2001) 769-774.
86. P.F. Biagi, R. Piccolo, A. Ermini, S. Martellucci, C. Bellecci, M. Hayakawa, V. Capozzi, S.P. Kingsley: "Possible Earthquake Precursors Revealed by LF Radio Signals" Natural Hazard and Earth System Sciences, 1, (2001) 99-104.
87. S. Pagliara, L. Sangaletti, L.E. Depero V. Capozzi, G. Perna: "Structural Disorder in CdS_xSe_{1-x} Films Probed by Microdiffraction Experiments" Applied Surface Science, 186, (2002) 527-532.
88. G. Perna, V. Capozzi, M.C. Plantamura, A. Minafra, P.F. Biagi, S. Orlando, V. Marotta, A. Giardini: "Structural and Optical Properties of Pulsed Laser-Deposited ZnSe films" Applied Surface Science, 186, (2002) 521-526.
89. G. Perna and V. Capozzi: "Analisys of Optical Spectra of CdS and CdSe Films Deposited on Sapphire Substrate by Laser Ablation Technique" European Physical Journal B 26, (2002) 401-406.
90. V.L. Berkovits, V.P. Ulin, M. Losurdo, P. Capezzuto, G. Bruno, G. Perna, V. Capozzi: "Wet Chemical Nitridation of GaAs(100) by Hydrazine Solution for Surface Passivation" Applied Phys. Letters 80, (2002) 3739-3741.
91. M. Losurdo, P. Capezzuto, G. Bruno, G. Perna, V. Capozzi: "N_2-H_2 Remote Nitridation for GaAs Surface Passivation" Applied Phys. Letters 81, (2002) 16-18.
92. G. Perna, V. Capozzi, M.C. Plantamura, A. Minafra, S. Orlando, V. Marotta: "ZnSe Films Deposited on Cristalline GaAs and Amorphous Quartz Substrates by means of Pulsed Laser Ablation Technique" European Physical Journal B 29, (2002) 541-545.
93. G. Perna, V. Capozzi, M.C. Plantamura, A. Minafra, P.F. Biagi, M. Pallara, S. Orlando, V. Marotta, A. Giardini: "Structural and Optical Properties of ZnSe Films Deposited on Crystalline Al_2O_3 Substrate by means of Pulsed Laser Ablation Technique" Applied Surface Science 208, (2003) 582-588.
94. V. Augelli, T. Ligonzo, A. Minafra, G. Perna, L. Schiavulli, V. Capozzi, M. Ambrico, M. Losurdo: "Optical and Electrical Characterization of n-GaAs surfaces Passivated by N_2-H_2 Plasma" J. Luminescence 102, (2003) 519-524.
95. P.F. Biagi, O. Molchanov, R. Piccolo, A. Minafra, A. Ermini, V. Capozzi, Y. M. Khatkevich: "Co-postseismic Hydrogeochemical Anomalies in a Volcanic Environement"Natural Hazard and Earth System Sciences, 3, (2003) 263-267.
96. P.F. Biagi, R. Piccolo, V.Capozzi, A. Ermini, S. Martellucci, C. Bellucci: "Exalting in Atmospheric Tides as Earthquake Precursor" Natural Hazard and Earth System Sciences, 3, (2003) 197-201.

97. G. Perna, V. Capozzi, A. Minafra, M. Pallara, M. Ambrico:"Effects on Indium Doping on Structural and Optical Properties of CdSe Thin Films Deposited by Laser Ablation Technique" European Physical Journal B 32, (2003) 339-344.
98. G. Perna, V. Capozzi, M. Ambrico, V. Augelli, T. Ligonzo, A. Minafra, L. Scihavulli, M. Pallara: "Structural and Optical Characterization of Undoped and Indium Doped CdS Films Grown by Pulsed Laser Deposition" Thin Solid Films 453, (2004) 187-194.
99. P.F. Biagi, R. Piccolo, A. Minafra, T. Maggipinto, L. Castellana, O. Molchalov, Ermini, V. Capozzi, G. Perna, Y.M. Khatkevich, E.I. Gordeev: "Retrospective Analysis for Detecting Seismic Precursors in Groundwater Argon Content" Natural Hazard and Earth System Sciences, 4, (2004) 9-15.
100. Vito Capozzi & Giuseppe Perna: "Structural and Optical properties of II-VI Semiconductor Thin Films Deposited by Laser Ablation "Recent Research Developments in Science & Technology of Semiconductors, vol. 1 (2003) 279-320; Ed. Transworld research Network, Kerala (India).
101. P.F. Biagi, R. Piccolo, A. Ermini, S. Martellucci, C. Bellecci, V. Capozzi, G. Perna, O. Molchalov, M. Hayakawa: "Variations in LF Radio Signal on the Occasion of the Recent Seismic and Volcanic Activity in Southern Italy" Physics and Chemistry of the Earth 29, (2004) 551-557.
102. G. Perna, V. Capozzi, M. Ambrico, V. Augelli, T. Ligonzo, A. Minafra, L. Scihavulli, M. Pallara:"Structural and Optical Characterization of Zn Doped CdS Films" Applied Surface Science 233, (2004) 366-372.
103. P.F. Biagi, L. Castellana, R. Piccolo, A. Minafra, G. Maggipinto, A. Ermini, V. Capozzi, G. Perna, Y.M. Khatkevitch, E.I. Gordeev: "Disturbances in Groundwater Chemical Parameter Related to Seismic and Volcanic Activityin Kamchatka (Russia)" Natural Hazard and Earth System Sciences, 4, (2004) 535-539.
104. Erlacher, M. Ambrico, V. Capozzi, V. Augelli, H. Jaeger, B. Ulrich "X-ray, Absorption and Photocurrent Properties of Thin Films GaAs on Glass Formed by Pulsed-laser Deposition" Semiconductor Science & Technology 19, (2004) 1322-1324.
105. V. Capozzi, G. Perna, G.F. Lorusso: "Photoluminescence Analysis of Si doped $Al_xGa_{1-x}As$ near the Mott Transition Doping Density" Recent Research Developments in Science & Technology of Semiconductors, vol. 2 (2004) 73-93; Ed. Transworld Research Network, Kerala (India).
106. P.F. Biagi, R. Piccolo, L. Castellana, T. Maggipinto, A. Ermini, S. Martellucci, C. Bellecci, V. Capozzi, G. Perna, O. A. Molchalov, M. Hayakawa, K. Ohta: "VLF-LF Radio Signals Collected at Bari (South Italy): a Preliminary Analysis on Signal Anomalies Associated with Earthquakes" Natural Hazard and Earth System Sciences, 4, (2004) 685-689.
107. G. Perna, A. Gallone, V. Capozzi, P.F. Biagi, A. Fratello, G. Guida, P. Zanna, R. Cicero: "Optical Spectra of Melanin Films Extracted from *Rana Esculenta*" Physica Scripta T118,(2005) 89-92
108. Piccoli, G. Perna, R. Scrima, O. Cela, R. Rinaldi, D. Boffoli, V. Capozzi, N. Capitanio:"A Novel Redox State Heme *a* Marker in Cytochrome c Oxidase Reveled by Raman Spectroscopy"Physica Scripta T118, (2005) 199-204.
109. V. Capozzi, G. Perna, A. Gallone, P.F. Biagi, P. Carmone, A. Fratello, G. Guida, P. Zanna, R. Cicero: "Raman and Optical Spectroscopy of Eumelanin Films" J. Molecular Structure 744, (2005) 717-721.

110. P.F. Biagi, L. Castellana, T. Maggipinto, R. Piccolo, A. Minafra, A. Ermini, S. Martellucci, C. Bellecci, G. Perna, V. Capozzi, O.A. Molchanov, M. Hayakawa: "A possible Preseismic Anomaly in the Ground Wave of a Radio Broadcasting (216 kHz) during July- August 1998 (Italy)" Natural Hazard and Earth System Sciences, 5, (2005) 727-732.
111. G. Perna, M. Lastella, M. Ambrico, V. Capozzi: "Temperature dependance of the Optical Properties of ZnSe Fims Deposited on Quartz Substrate" Applied Physics A 83, (2006) 127-130.
112. V. Capozzi, G. Perna, P. Carmone, A. Gallone, M. Lastella, E. Mezzenga, G. Quartucci, M. Ambrico, V. Augelli, P.F. Biagi, T. Ligonzo, A. Minafra, L. Schiavulli, M. Pallara, R. Cicero: "Optical and Photoelectronic Properties of Melanin" Thin Solid Films 511, (2006) 362-366.
113. P.F. Biagi, L. Castellana, T. Maggipinto, R. Piccolo, A. Minafra, A. Ermini, S. Martellucci, C. Bellocci, G. Perna, V. Capozzi, O.A. Molchanov, M. Hayakawa: "LF radio anomalies revealed in Italy by wavelet analysis: Possibile preseismic effects during 1997-1998. Physics and Chemistry of the Earth, 31, (2006) 403-408.
114. V. Capozzi, G. Perna, S. Pagliara, L. Sangaletti: "Structural Disorder in CdS_xSe_{1-x} Thin Films Probed by Microdiffraction and Optical Spectroscopies" New Research on Semiconductors vol. 1, ch.6, (2006) 143-157; Ed. T.B. Elliot, Nova Science Publishers, (USA).
115. P.F. Biagi, L. Castellana, T. Maggipinto, A. Ermini, G. Perna, V. Capozzi: "Electric Field Strength Analysisof 216 and 270 kHz Broadcast Signals Recorded During 9 Years" Radio Science 41, (2006) 1-11.
116. Vito Capozzi, Giuseppe Perna: "Role of Deposition Parameters and Aging on the Optical and Photoluminescence Properties of C_{70} Films" Thin Solids Films 515, (2007) 7247-7252.
117. G. Perna, M. Lastella, M. Lasalvia, E. Mezzenga, V. Capozzi: "Raman Spectroscopy and Atomic Force Microscopy Study of Cellular Damage in Human keratinocytes Treated with $HgCl_2$" J. Molecular Structure 834, (2007) 182-187.
118. M. Lastella, M. La salvia, G. Perna, PF. Biagi, V. Capozzi: "Atomic Force Microscopi Study on Human Keratinocytes treated with $HgCl_2$" J. Physics: Conference Series 61, (2007) 920-925.
119. Gallone, A. Sagliano, G. Guida, S. Ito, K. Wakamatsu, V. Capozzi, G. Perna, P. Zanna, R.Cicero: "The Melanogenic System of the Liver Pigmented Macrophages of Rana Esculenta:"Tyrosinase Activity Histology and Histoptology 22, (2007) 1065-1075.
120. M. Buttiglione, L. Roca, E. Montemurro, F. Vitello, V. Capozzi, G. Cibelli: "Radiofrequency Radiation (900 MHz) induces Egr-1 Gene Expression and affects Cell-cycle control Neuroblastoma Cells"J. Cell Biology 213, (2007) 759-767.
121. P.F. Biagi, L. Castallana, T. Maggipinto, G. Maggipinto, A. Ermini, V. Capozzi, G. Perna, M. Solovieva, A. Rozhnoi, O.A. Molchanov, M. Hayakawa: "Decrease in the Electric intensity of VLF/LF Radio Signals and possible connections" Natural Hazard and Earth System Sciences, 7, (2007) 423-430.
122. M.C. Frassanito, C. Piccoli, V. Capozzi, D. Boffoli, A. Tabilio, N. Capitanio: "Topological Organization of NADPH-Oxidase on Hematopoietic Stem Cell Membrane: Preliminary Studyby Fluorescence Near Field Optical Microroscopy" J. Microscopy, 229, (2008) 517-524.

123. P.F. Biagi, L. Castallana, T. Maggipinto, D. Loiacono, L. Schiavulli, A. Ermini, V. Capozzi, M.S. Solovieva, A.A. Rozhonoi. O.A. Molchanov, M. Hayakawa: "Disturbances in a VLF Radio Signal prior the M=4.7 offshore Anzio (central Italy) Earthquake on 22 August 2005" Natural Hazard and Earth System Sciences, 8, (2008) 1041-1048.
124. Camerlengo, F. Zenone, G. Perna, V. Capozzi, N. Cirillo, G. M. Gaeta, M. Lepore: "Investigation on Micro-Raman Spectra and Wavelet Data Analysis for Pemphigus Vulgaris Follow-up Monitorino" Sensors, 8, (2008) 3656-3664.
125. G. Perna, M.C. Frassanito, G. Palazzo, A. Maliardi, P.F. Biagi, V. Capozzi: "Florescence Spectroscopy of Synthetic Melanin in Solution" J. Luminescence, 129, (2009) 44-49.
126. T. Ligonzo, M. Ambrigo, V. Augelli, G. Perna, L. Schiavulli, M.A. Tamma, P.F. Biagi, Minafra, V. Capozzi: "Electrical and Optical Properpies of Natural and Synthetic Melanin Biopolymer" J. Non-Crystalline Solids, 355, (2009) 1221-1226.
127. P.F. Biagi, L. Castellana, T. Maggipinto, G. Maggipinto, T. Ligonzo, l. Schiavulli, D. Loiacono, A. Ermini, M. Lasalvia, G. Perna, V. Capozzi: "A Reverberation Chamber to Investigate Possibile Effects of *In Vivo* Exposure of Rats to 1.8 GHz Electromagnetic Fields: A Preliminary Study" Progress in Electromagnetics Research, 94, (2009) 133-152.
128. G.Perna, M. Lasalvia, A. Castro, E. Mezzenga, N. L'Abbate, P.F. Biagi, V. Capozzi: "Detection of Pestcide Effects in Human Keratinocytes by means of Raman Microspectroscopy" Applied Physics Letters, 95, (2009) 083701.
129. G.Cicala, R. Brescia, M.A. Nitti, A. Romeo, M. Ambrico, L. Schiavulli, G. Perna, V. Capozzi: "Thermoluminescent Response of Thin (2μm) Polycristalline Diamond Films Grown by Pulsed and Continuous Microwave Plasmas" Diamond Related Materials, 19, (2010) 470-473.
130. M. Ambrico, A. Cardone, T. Ligonzo, V. Augelli, P.F. Ambrico, S. Cicco, G.M. Farinosa, M. Filannino, G. Perna, V. Capozzi: "Hysteresis-type current-voltage characteristics in Au/eumelanin/ITO/glass structure: Towards melanin based memory devices" Organic Electronics, 11, (2010) 1809-1814.
131. G. Perna, M. Lasalvia, P. D'Antonio, N. L'Abbate, V. Capozzi: "Identification of Chemical Modification in Single Human Keratinocytes Cells Exposed to low Doses of Chlorpiriphos by Raman Micro-spectrscopy" J. Raman Spectroscopy 42, (2011) 603-611.
132. M. Lasalvia, G. Perna, E. Mezzenga, E. Migliorini, M. Lazzarino, N. L'Abbate, V. Capozzi: "Atomic Force Microscopy Investgation of morphological changes in Living Keratinocytes Treated with $HgCl_2$ at not cytotossic Doses"J. Microscopy, (2011).(in press)
133. G. Perna, G. Palazzo, A Mallardi, V. Capozzi: "Fluorescence Properties of Natural Eumelanin Biopolymer"J. Luminescence, (2011) (in press)

Vito CAPOZZI: PUBBLICATIONS IN PROCEEDINGS OF INTERNATIONAL CONFERENCES (peer reviwed)

1. P1- JL. Staehli and V. Capozzi:
"Coexistence of Direct and Indirect E-H Plasma Components in GaSe"
Proceedings of the "17th Inter. Confer. on the Physics of Semiconductors",
San Francisco. (USA), 1984; Ed. J.D.Chadi and W.A. Harrison,
Springer-Verlag, pp. 1297-1300
2. P2- V. Capozzi, K. Maschke, M. Montagna:
"Influence of Stacking Disorder on the Photoluminescence of GaSe"
Proceedings of the "18th Inter. Confer. on the Physics of Semiconductors",
Stockholm (Sweden), 1986; Ed. O.Engstrom, World Scientific, pp.1429-1432
3. P3- V. Capozzi, L. Colangeli, E. Bussoletti, A. Minafra, A. Borghesi:
"Size Effect in Optical Spectra of Amorphous Carbon Grains"
Proc. of the "Synchrotron Radiation at Frascati: 1986 Users Meeting",
Frascati (Italy); Ed.. S. Mobilio et al, Italian Physical Society,
Conf. Proceed. 5, (1986) 159-163.
4. P4- V. Capozzi, C. Flesia, A. Minafra:
"Light Standing-Waves in Aggregates of Amorphous Carbon Grains"
Proc. of the "Int. Conf. on Experiments on Cosmic Dust Analogues",
Capri (Italy) 1987; edits.E. Bussoletti et al., Kluwer Acad. Publ. pp. 153-160.
5. P5- V. Capozzi, JL. Staehli, C. Flesia, D. Martin, V. Augelli, GF. Lorusso:
"Exciton Effects in Absorption Spectra of GaAs/AlGaAs Superlattices
Proceed. of the Inter. Conf. on "Optics of Excitons in Confined Systems",
Giardini Naxos (Italy) 1991; Ed. A. Quattropani et al.; Inst. of Physics,
Conference Series 123, (1992) 195-198.
6. P6- G.F. Lorusso, V. Capozzi, A. Minafra:
"Study of the Noise of Correlation and Structure Functions".
Proceed. of the Inter. Conference on "Photon Correlation andScattering:
Theory and Applications", Boulder (USA), August 1992;Ed. Technical Digest Series 20, (1992) 61-63.
7. P7- N. Senesi, G.F. Lorusso, T.M. Miano, G. Maggipinto, F. Rizzi, V. Capozzi:
"The Fractal Dimension of Humic Substances as a Function of pH by
Turbidity Measurements"
Proc. of the Int. Conf. on "Humic Substances in Global Environment and
Implications on Human Health" Bari (Italy) 1993,
Ed. N. Senesi and T.M. Miano, Elsevier Science (1994), pp. 121-126.
8. P8- V. Capozzi, C. Flesia, G.F. Lorusso, D. Martin, G. Perna, JL. Staehli:
"Optical Spectroscopy of GaAs/AlGaAs Superlattices with Randomly
Distributed Well Widths"
Proc. of the Intern. Conf. on "Semiconductor Heteroepitaxy", Montpellier (F)
1995; Ed. B. Gil and R.L. Aulombard, World Scientific, pp. 494-497;
9. P9- M. Ambrico, D. Smaldone, R. Martino, V. Capozzi, G. Perna, G.F. Lorusso, Giardini, A. Mele:
"Photoluminescence and X-ray Analysis of Laser-Deposited II-VI Thin
Films and Multilayered Structures"

Proc. Of the "1996 Summer Topical Meetings on Advanced Applications of Lasers in Material Processing", Keystone (USA) 1996; IEEE/LEOS, No. 96TH8164, pp. 19-20.
10. P10- M. Ambrico, R. Martino, D. Smaldone, V. Capozzi, G.F. Lorusso, G. Perna, A. Giardini, A. Mele: "Laser Ablation of Highly Oriented CdSe Thin Films and CdSe/CdTe Multilayers on Silicon Substrate" Proc. of the Symposium on "Adv. Laser Processing of Materials: Fundamentals and Applications"; 1995 MRS Fall Meeting, Boston (USA) 1995; Eds. Rajiv Singh et al., Vol. 397 (1996) 125-130.
11. P11- G. Cicala, G. Bruno, P. Capezzuto, L. Schiavulli, V. Capozzi, G Perna: "Deposition of Photoluminescent Nanocrystalline Silicon Films by SiF4-SiH4-H2 Plasmas". Proc. of Material Research Soc., Fall. Meeting, Boston (USA) 1997; Vol. 452 (1997) 809-814.
12. P12- G. Perna, V. Capozzi, M. Ambrico, D. Smaldone, R. Martino: "Laser Ablation and Deposition of CdSe and CdS on GaAs Substrate" Proc. of the Intern. Conf. on "Laser Surface Processing", Limoges (F) 1997; Ed. V.I. Pustovoy, Proc. SPIE vol. 3404, (1998) 33-38.
13. P13- M. Ambrico, V. Stagno, D. Smaldone, R. Martino, G. Perna, V. Capozzi: "Pulsed Laser Deposition, of Zn_xSe_{1-x} Thin Films on Quartz: Measurement of Energy Gap, Absorption Coefficient and Refractive Index" Proc. of the Intern. Conf. on "Laser Surface Processing", Limoges (F) 1997; Ed. V.I. Pustovoy, Proc. SPIE vol. 3404, (1998) 39-46.
14. P14- G. Perna, V. Capozzi, S. Pagliara, M. Ambrico: "Temperature Study of Absorption and Photoluminescence Spectra of CdS_xSe_{1-x} Films Grown on Quartz Substrate by Pulse Laser Ablation Technique" Proc. of the Intern. Conf. on "Advanced Laser Technologies", Lecce (I) 1999; Eds. V.I. Pustovoy & V. I. Konov, Proc. SPIE vol. 4070, (2000) 372-379.
15. P15- F. Zenone, M. Lepore, G. Perna, P. Carmone, R. Riccio, G.M. Gaeta, V. Capozzi: "Micro-Raman Spectroscopy on Oral Tissues" Proc. of the "XII Intern. Conf. on Lasers in Dentistry", Eds. P. Rechmann & D. Fried, Proc. SPIE vol. 6137, (2006) 61370P 1-9.
16. P16 - F. Zenone, M. Lepore, G. Perna, P. Carmone, I. Delfino, G.M. Gaeta, V. Capozzi: "Oral Patology Diagnosis by means of Micro-Raman spectroscopy on Biopsies and Blood serum" Proc. of the "XIII Intern. Conf. on Lasers in Dentistry", Eds. P. Rechmann & D. Fried, Proc. SPIE vol. 6425, (2007) 64250V 1-11.
17. P17 - C. Camerlingo, F. Zenone, G. Perna, V. Capozzi, N. Cirillo, G.M. Gaeta, M. Lepore: "Wavelet data analysis of micro-Raman spectra for follow-up monitoring in oral pathologies" Proc. of the "XIV Intern. Conf. on Lasers in Dentistry", Eds. P. Rechmann & D. Fried, Proc. SPIE vol. 6843, (2008) 68430K 1-10.
18. P18 - V. Petruzzelli, G. Calò, A. D'Orazio, M. De Sario, L. Mescia, F. Prudenzano V. Capozzi, L. Castellana, T. Maggipinto, P.F. Biagi:

"Reverberation Chambers per Electromagnetic Fields"
Proc. of the "5th Intern. Workshop on Biological Effects of Electromagnetic Fields", Terrasini, Palermo (Italy) 2008.

19. P19 - I. Delfino, C. Camerlengo, F. Zenone, G. Perna, V. Capozzi, N. Cirillo, G.M. Gaeta, E. De Mol, M. Lepore:
"Oral Patology Follow-up by means of Micro-Raman Spectroscopy on Tissue and Blood Serum Samples: an Application of Wavelet & Multivariate DataAnalysis"
Proc. of the "XV Intern. Conf. on Lasers in Dentistry",
Eds. P. Rechmann & D. Fried, Proc. SPIE vol. 7162, (2009) 71620L 1-10.
20. P20 - I. Delfino, C. Camerlengo, F. Zenone, G. Perna, V. Capozzi, N. Cirillo, G.M. Gaeta, M. Lepore:
"Oral Pathology Follow-up by means of Micro-Raman Spectroscopy on Tissue and Blood Serum Samples: an Application of Wavelet and Multivariate Data Analysis"
DGaO Proceedings 2009
21. P21 - I. Delfino, C. Camerlengo, F. Zenone, G. Perna, V. Capozzi, N. Cirillo, G.M. Gaeta, M. Lepore:
Micro-Raman Spectroscopy of Tissue Samples for Oral Pathology Follow-up Monitoring"
Proc. of "Biophotonics: Photonic Solutions for Better Health Care II"
Ed. J. Popp, W. Drexler, V.V. Tuchin & D.L. Matthews, Proc. SPIE vol. 7715, (2010) 77152J 1-8.
22. P22 - C. Camerlengo, I. Delfino, G. Perna, V. Capozzi, M. Lepore:
"Micro-Raman Spectroscopy and Univariate Analysis for Medical Diagnosis"
XII International Conference on Raman Spectroscopy, Boston (USA) 2010,
Eds. P.M. Champion & L. D. Ziegler, AIP Conference Proceedings, Vol. 1267 (2010) 384-385.

In: Chemistry Researcher Biographical Sketches
Editors: E. S. Huseynov and V. Babayev
ISBN: 978-1-62100-771-5

Chapter 11

BIOGRAPHICAL SKETCH

Name:	Title:	Date of Birth:
Igor Čeliković	Institute of Nuclear Sciences "Vinča	21.02.1978

EDUCATION:

Institution and Location	Degree	Year Conferred	Scientific Field
University of Belgrade , Faculty of Electrical Engineering, Department of Nuclear and Medicine Techniques		2002	
University of Belgrade, Faculty of Physics, Magisterium Thesis Degree "Optimisation of measurement of enhanced radon concentration"		2006	

CONTACT POINTS:

Mike Petrovića Alasa 12-14, 11000 Belgrade, Serbia

RESEARCH AND PROFESSIONAL EXPERIENCE:

- 2003 - Institute of Nuclear Sciences "Vinča"
- Participation in the Project Nuclear Physics, Methods and Instruments (2003-)

- Participation in bilateral Slovenian-Serbian project "Population exposure to radon and thoron in living and working environment in Slovenia and Serbia and Montenegro"
- Participation in EC FP6 project: INTAILRISK (Assessment Of Environmental Risk Of Radioactively Contaminated Industrial Tailings), Contract No INCO-CT-2004-509214
- Participation in EC FP6 project: INDUWASTE (Management and Remediation of Hazardous Industrial Wastes in the Western Balkan Countries), Contract N° 515919
- 5 th International Balkan School in Nuclear Physics, "Physics with Radioactive Beams", 7-14 September 2006, Brasov , Romania
- Participation in bilateral French-Serbian CNRS/MSCI collaboration agreement (No 20505 "Towards SPIRAL 2 and EURISOL")

Publications during last three years

1. G. Milić, B. Jakupi, S. Tokonami, R. Trajković, T. Ishikawa, I. Čeliković, P. Ujić, O. Čuknić, I.V. Yarmoshenko, K. Kosanović, F. Adrović, S.K. Sahoo, N. Veselinovic, Z.S. Žunić, The concentrations and exposure doses of radon and thoron in residences of the rural areas of Kosovo and Metohija, Radiation Measurements, Volume 45, Issue 1, January 2010, Pages 118-121
2. Predrag Ujić, Igor Čeliković, Aleksandar Kandić, Ivana Vukanac, Mirjana Đurašević, Dušan Dragosavac, Zora S. Žunić; Internal exposure from bulding materials exhaling 222Rn and 220Rn as compared to external exposure due to their natural radioactivity content, Applied Radiation and Isotopes, *Applied Radiation and Isotopes*, Volume 68, Issue 1, January 2010, Pages 201-206
3. Field experience with soil gas mapping using Japanese passive radon/thoron discriminative detectors for comparing high and low radiation areas in Serbia (Balkan Region), Zunic ZS , Janik M , Tokonami S , Veselinovic N , Yarmoshenko IV , Zhukovsky M , Ishikawa T , Ramola RC , Ciotoli G , Jovanovic P , Kozak K , Mazur J , Celikovic I , Ujic P , Onischenko A, Sahoo S, Bochicchio F, Journal of radiation research, 50, 2009 355-361

In: Chemistry Researcher Biographical Sketches
Editors: E. S. Huseynov and V. Babayev
ISBN: 978-1-62100-771-5

Chapter 12

BIOGRAPHICAL SKETCH

Name:	Title:	Date of Birth:
Monica C. Chuong	Massachusetts College of Pharmacy and Health Sciences	07-05-1954

EDUCATION:

Institution and Location	Degree	Year Conferred	Scientific Field
University of Houston, Oregon State University, Postdoctorate	PhD;		

CONTACT POINTS:

179 Longwood Ave, W316, Boston, MA 02115

Awards and Honors:

2002 - Goel Academic Achievement Award, University of Houston

RESEARCH AND PROFESSIONAL EXPERIENCE:

Preformulation, high performance liquid chromatography, solid dosage formulations, in vitro drug release technologies, drug stability

Professional Appointments:

Assistant Professor of Pharmaceutics

Publications during last three years

1. Christensen JM, ***Chuong MC***, Le H, Pham L, and Bendas E. Hydrocortisone Diffusion Through Synthetic Membrane, Mouse Skin, and Epiderm™ Culture Skin. Arch Drug Information, 4:10-21, 2011.
2. ***Chuong MC***, Prasad D, LeDuc B, Du B, Putcha L. Stability of Vitamin B Complex in Multivitamin and Multimineral Supplement Tablets after Space Flight. Journal of Pharmaceutical and Biochemical Analysis, 55: 1197-1200, 2011.
3. ***Chuong MC***, Palugan L, Su T, Lee R, Busano C, De Pretoro G, Shah A. Formulation of Controlled Release Capsule of a Biopharmaceutical Classification System I Drug Using Niacin as a Model. AAPSPharmSciTech, 11(4): 1650-61, 2010.
4. ***Chuong MC***, Taglieri C, Crosby S, Ferullo J and Wei P. Effect of Beverage on the In Vitro Disintegration of Immediate Release Pain Medications. Dissolution Technologies, 17(1) 31-37, 2010.
5. ***Chuong MC***, Lee RG, Saxena V and Palugan L. USP Disintegration Apparatus as a Potential Tool for Evaluating Drug Release from Controlled-Release Dosage Forms. Dissolution Technologies, 16(1) 25-32, 2009.
6. ***Chuong MC***, Christensen JM, Ayres JW. New Dissolution Method for Mesalamine Tablets and Capsules. Dissolution Technologies, 15 (3) 7-14, 2008.
7. ***Chuong MC***, Christensen JM, Ayres JW. Sustained Delivery of Intact Drug to the Colon: Mesalamine Formulation and Temporal Gastrointestinal Transit Analysis. Pharmaceutical Development and Technology (Pharm Dev Technol), 27:1-10, 2008.

In: Chemistry Researcher Biographical Sketches
Editors: E. S. Huseynov and V. Babayev

ISBN: 978-1-62100-771-5

Chapter 13

BIOGRAPHICAL SKETCH

Name:	Title:	Date of Birth:
Maria Vittoria Diamanti	Ph.D	August 04th 1981

EDUCATION:

Institution and Location	Degree	Year Conferred	Scientific Field
M.D. in Materials Engineering at Politecnico di Milano.		2005:	
Rocca Fellow at MIT as visiting student (1 semester). 2009: Ph.D. in Materials Engineering at Politecnico di Milano		2007	

CONTACT POINTS:

Email: mariavittoria.diamanti@polimi.it
telephone: +39 2 23993137

RESEARCH AND PROFESSIONAL EXPERIENCE:

Since May 2009: Assistant Professor at Politecnico di Milano. Research activity mainly concerning titanium dioxide, from an electrochemical, biomedical, photocatalytic, and aesthetical point of view. Teaching assistant in both Engineering and Architecture classes dealing with materials, corrosion of metals and materials durability. Member of the Technical Committee of the Clean Technology Conference and Expo (since 2008). Scientific secretary of the annual PhD School organised by AIMAT – Italian Association of Materials Engineering (since 2008).

Publications during last three years

1. B. Del Curto, M.V. Diamanti, P. Dalla Pria, F. Sbaiz, A. Cigada, Anodic Spark Deposition treatments to increase reliability of Ti6Al4V modular prostheses, Journal of Applied Biomaterials & Biomechanics (2009), vol. 7, pp. 153-159.
2. Maria Vittoria Diamanti, Barbara Del Curto, Alberto Barlattani, Patrizio Bollero, Liliana Ottria, MariaPia Pedeferri, Mechanical characterization of an innovative dental implant system, Journal of Applied Biomaterials & Biomechanics (2009), vol. 7, pp. 23-28.
3. M.V. Diamanti , S. Codeluppi, A. Cordioli, MP. Pedeferri, Effect of thermal oxidation on titanium oxides characteristics, J. Experimental Nanoscience 4 (2009), 365-372.
4. M.V. Diamanti, M. Ormellese, MP. Pedeferri, Photoactivity of nitrogen-doped anodic TiO_2 under UV and VIS light irradiation – Nanotech 2009 (Houston, TX, USA), Technical Proceedings of the 2009 Nanotechnology Conference and Trade Show, vol. 3, pp. 198-201, ISBN: 978-1-4398-1784-1.
5. M.V. Diamanti, M. Ormellese, MP. Pedeferri, Tuning of titanium oxide morphology at micro and nano scale by alternating current anodizing, J. Nano Research (2009), vol. 6, pp. 61-66.
6. M.V. Diamanti, MP. Pedeferri, C. A. Schuh, Thickness of anodic titanium oxides as a function of crystallographic orientation of the substrate, Metallurgical and Materials Transactions A (2008), vol. 39A, pp. 2143-2147
7. M.V. Diamanti, M. Ormellese, MP. Pedeferri, Characterization of photocatalytic and superhydrophilic properties of mortars containing titanium dioxide, Cement and Concrete Research (2008) vol. 38, pp. 1349-1353.
8. M. V. Diamanti, B. Del Curto, MP. Pedeferri, Interference colors of thin oxide layers on titanium, Color Research and Application (2008), vol. 33, pp. 221-228.
9. M.V. Diamanti, B. Del Curto, M. Ormellese, MP. Pedeferri, Photoactive and coloured anodic oxides on titanium for architectural and design applications – Cleantech 2008 (Boston, MA, USA), Clean Technology 2008: Bio Energy, Renewables, Green Building, Smart Grid, Storage, and Water, vol. 1, pp. 170-173, ISBN: 1-4200-8502-0.
10. M.F. Brunella, M.V. Diamanti, M.P. Pedeferri, F. Di Fonzo, C.S. Casari, A. Li Bassi, Photocatalytic behaviour of different titanium dioxide layers, Thin Solid Films (2007) 515, pp. 6309-6313.
11. Maria Vittoria Diamanti, MariaPia Pedeferri, Effect of anodic oxidation parameters on the titanium oxides formation, Corrosion Science (2007), vol. 49, pp. 939-948.

In: Chemistry Researcher Biographical Sketches
Editors: E. S. Huseynov and V. Babayev
ISBN: 978-1-62100-771-5

Chapter 14

BIOGRAPHICAL SKETCH

Name:	Title:	Date of Birth:
Sergey V. Dorozhkin	Ph.D. in chemistry	June 24, 1961

EDUCATION:

Institution and Location	Degree	Year Conferred	Scientific Field
	Ph.D. in chemistry		

CONTACT POINTS:

Kudrinskaja sq. 1 - 155, 123242 Moscow D-242, Russia

AWARDS AND HONORS: N/A

RESEARCH AND PROFESSIONAL EXPERIENCE:

A number of post-doctoral positions in calcium phosphates, bioceramics, biomaterials and biomineralization

Professional Appointments:

N/A

Publications during last three years

1. Dorozhkin S.V.: A novel, environmentally friendly process for the fabrication of calcium phosphate bioceramics. Inorg. Mater. 2008, 44 (2), 207 – 210.
2. Dorozhkin S.V.: Calcium orthophosphate cements for biomedical application. J. Mater. Sci. 2008, 43 (9), 3028 – 3057.
3. Dorozhkin S.V.: Green chemical synthesis of calcium phosphate bioceramics. J. Applied Biomaterials & Biomechanics 2008, 6 (2), 104 – 109.
4. Dorozhkin S.V.: Calcium orthophosphate cements and concretes. Materials 2009, 2 (1), 221 – 291.
5. Dorozhkin S.V.: Calcium orthophosphate-based biocomposites and hybrid biomaterials. J. Mater. Sci. 2009, 44 (9), 2343 – 2387.
6. Dorozhkin S.V.: Calcium orthophosphates in nature, biology and medicine. Materials 2009, 2 (2), 399 – 498.
7. Dorozhkin S.V., Ajaal T.: Toughening of porous bioceramic scaffolds by bioresorbable polymeric coatings. Proc. I. Mech. Eng. Part H 2009, 223 (H4), 459 – 470.
8. Dorozhkin S.V.: Nano-sized and nanocrystalline calcium orthophosphates in biomedical engineering. J. Biomimetics, Biomaterials and Tissue Engineering 2009, 3, 59 – 92.
9. Dorozhkin S.V., Ajaal T.: Strengthening of dense bioceramic samples using bioresorbable polymers – a statistical approach. J. Biomimetics, Biomaterials and Tissue Engineering 2009, 4, 27 – 39.
10. Dorozhkin S.V.: Nanodimensional and nanocrystalline apatites and other calcium orthophosphates in biomedical engineering, biology and medicine. Materials 2009, 2 (4), 1975 – 2045.
11. Dorozhkin S.V.: Nanosized and nanocrystalline calcium orthophosphates. Acta Biomater. 2010, 6 (3), 715 – 734.
12. Jin H.B., Guo C.B., Mao K.Y., Dorozhkin S.V., Agathopoulos S.: Preparation of porous biphasic -TCP/HA bioceramics with a natural trabecular structure from calcined cancellous bovine bone. J. Ceram. Soc. Jpn. 2010, 118 (1), 52 – 56.

ISBN: 978-1-62100-771-5

Chapter 15

BIOGRAPHICAL SKETCH

Name:	Title:	Date of Birth:
Snežana Dragović	University of Belgrade, Institute for the Application of Nuclear Energy	July 27, 1971

EDUCATION:

Institution and Location	Degree	Year Conferred	Scientific Field
	PhD in Physical Chemistry		

CONTACT POINTS:

Banatska 31b, 11080 Belgrade, Serbia

RESEARCH AND PROFESSIONAL EXPERIENCE:

13 years experience in radioecology and heavy metal chemistry

Professional Appointments:

Geological Institute of Serbia (1998-2000), Institute for the Application of Nuclear Energy (2000-)

Publications during last three years

S. Dragović, O. Nedić, Bioindicators in the Assessment of Atmospheric Radioactivity: Current Approaches and Perspectives, In: Radioactive Contamination Research Developments, N.K. Henshaw, C.S. Alleyne (Ed.), Nova Science Publishers, Inc., ISBN 978-1-60741-174-1, New York, USA, 2010, pp. 43-66.

M. Momčilović, J. Kovačević, S. Dragović, Population doses from terrestrial exposure in the vicinity of abandoned uranium mines in Serbia, *Radiat. Meas.* 45 (2010) 225-230.

S. Dragović, Lj. Janković Mandić, Transfer of radionuclides to ants, mosses and lichens in semi-natural ecosystems", Radiat. Environ. Biophys. 46 (2010) 625-634.

S. Dragović, B.J. Howard, J.A. Caborn, C.L. Barnett, N. Mihailović, Transfer of natural and anthropogenic radionuclides to ants, bryophytes and lichen in a semi-natural ecosystem, *Environ. Monit. Assess.* 166 (2010) 677-686.

S. Dragović, N. Mihailović, B. Gajić, Quantification of transfer of ^{238}U, ^{226}Ra, ^{232}Th, ^{40}K and ^{137}Cs in mosses of a semi-natural ecosystem, *J. Environ. Radioact.* 101 (2010) 159-164.

Lj. Janković Mandić, R. Dragović, S. Dragović, Distribution of lithogenic radionuclides in soils of the Belgrade region (Serbia), *J. Geochem. Explor.* 105 (2010) 43-49.

Lj. Janković Mandić, S. Dragović, Assessment of terrestrial gamma exposure to the population of Belgrade (Serbia), *Radiat. Prot. Dosim.* 140 (2010) 369-377.

S. Dragović, N. Mihailović, Analysis of mosses and topsoils for detecting sources of heavy metal pollution: multivariate and enrichment factor analysis, *Environ. Monit. Assess.* 157 (2009) 383-390.

In: Chemistry Researcher Biographical Sketches
Editors: E. S. Huseynov and V. Babayev
ISBN: 978-1-62100-771-5

Chapter 16

BIOGRAPHICAL SKETCH

Name:	Title:	Date of Birth:
Richard Drevet	INSERM / URCA	November 03rd, 1980

EDUCATION:

Institution and Location	Degree	Year Conferred	Scientific Field
Ph.D in Materials Science			

CONTACT POINTS:

21, rue Clément ADER; BP 138; F-51685 Reims CEDEX 2

RESEARCH AND PROFESSIONAL EXPERIENCE:

Calcium phosphate coating,
Biomaterials, electrochemistry, Materials science

In: Chemistry Researcher Biographical Sketches
Editors: E. S. Huseynov and V. Babayev

ISBN: 978-1-62100-771-5

Chapter 17

BIOGRAPHICAL SKETCH

Name:	Title:	Date of Birth:
Barbora Drtinova	MSc. Ph.D.	1976/02/21

EDUCATION:

Institution and Location	Degree	Year Conferred	Scientific Field
Dept. of Nuclear Chemistry, Faculty of Nuclear Science and Physical Engineering, Czech Technical University, Brehova 7, 115 19 Prague 1, Czech Republic			
Nuclear chemistry at Dept. of Nuclear Chemistry, Faculty of Nuclear Science and Physical Engineering, Czech Technical University in Prague	Ph.D.	2007	

CONTACT POINTS:

Waldhauserova 140, 25228 Cernosice, Czech Republic

AWARDS AND HONORS:

Co-chairman at RadChem 2010

RESEARCH AND PROFESSIONAL EXPERIENCE:

Since February 2006 researcher and teaching assistant at CTU

Publications during last three years

Drtinova, B., Pospisil, M.: Behavior of cadmium (II) in irradiated aqueous solutions. *Journal of Radioanalytical and Nuclear Chemistry*. 2010, vol. 283, no. 3, p. 631-636. ISSN 0236-5731.

Drtinova, B. Pospisil, M., Cuba, V.: Products of radiation removal of lead from aqueous solutions. *Applied Radiation and Isotopes* . 2010, vol. 68, no.4-5, p.672-675. ISSN 0969-8043.

Vopalka, D., Stamberg, K., Motl, A., Drtinova, B.: The study of the speciation of uranyl-sulphate complexes by UV-Vis absorption spectra decomposition. *Journal of Radioanalytical and Nuclear Chemistry* . 2010, vol. 286, no. 3, p. 681-686. ISSN 0236-5731.

In: Chemistry Researcher Biographical Sketches
Editors: E. S. Huseynov and V. Babayev

ISBN: 978-1-62100-771-5

Chapter 18

BIOGRAPHICAL SKETCH

Name:	Title:	Date of Birth:
Pavel N. D'yachkov	Professor at Quantum Chemistry Laboratory, Kurnakov Institute of General and Inorganic Chemistry, Russian Academy of Sciences (IGIC RAS).	02.12.1947

EDUCATION:

Institution and Location	Degree	Year Conferred	Scientific Field
MS degree in chemistry from the University of Sankt-Petersburg (the former Leningrad)		1973	
Cand. Sci. (PhD) and Dr. Sci. degrees in physical chemistry from the IGIC RAS and Semenov Institute of Chemical Physics RAS		1978 and 1989.	

CONTACT POINTS:

Leninskii pr. 31, Moscow 119991, Russian Federation

RESEARCH AND PROFESSIONAL EXPERIENCE:

He is the author of more than 100 journal papers and has written five books including "Heteroligand Molecules: Structure, Isomers, Transformations" (Francis and Taylor, London and New York), 2002, 271 p; with A.A.Levin). His current research activity lies in quantum chemistry of the coordination compounds and nanomaterials, as well as in modeling of the toxic and carcinogenic effects of environmental pollutants.

Professional Appointments:

Leading investigator

Publications during last three years

1. P.N. D'yachkov, N. V. Kharchevnikova, Z. I. Zholdakova, et al. Quantum Chemical Metabolism-Based Simulation of Carcinogenic Potency of Benzene Derivatives. Intern. J. of Quantum Chem. 110, 1402–1411 (2010).
2. P. N. D'yachkov, D. Z. Kutlubaev, and D. V. Makaev. Linear augmented cylindrical wave Green's function method for electronic structure of nanotubes with substitutional impurities. PHYSICAL REVIEW B 82, 035426 (2010) (15 pages).
3. A. M. Nemilentsau, M. V. Shuba, G. Ya. Slepyan, P. P. Kuzhir, and S. A. Maksimenko, P. N. D'yachkov, andA. Lakhtakia. Substitutional doping of carbon nanotubes to control their electromagnetic characteristics. PHYSICAL REVIEW B 82, 235424 (2010) (10 pages).
4. P. N. D'yachkov and D. V. Makaev, Electronic structure of BN nanotubes with intrinsic defects N_B and B_N and isoelectronic substitutional impurities P_N, As_N, Sb_N, In_B, Ga_B, and Al_B. J. Phys. Chem. Solids. 70, 180-185 (2009).
5. P. N. D'yachkov and D. V. Makaev, Description of band structures of armchair nanotubes using the symmetry-adapted linear augmented cylindrical wave method, Phys. stat. solidi. (b) 246, 140-146 (2009).

In: Chemistry Researcher Biographical Sketches
Editors: E. S. Huseynov and V. Babayev
ISBN: 978-1-62100-771-5

Chapter 19

BIOGRAPHICAL SKETCH

Name:	Title:	Date of Birth:
Fernando S. García Einschlag	Instituto de Investigaciones Fisicoquímicas Teóricas y Aplicadas (INIFTA), CCT-La Plata-CONICET, Departamento de Química, Facultad de Ciencias Exactas, Universidad Nacional de La Plata., Argentina.	26-03-1969

EDUCATION:

Institution and Location	Degree	Year Conferred	Scientific Field
PhD, University of La Plata, Argentina			

CONTACT POINTS:

64 Diag. 113, CP (1900), La Plata, Buenos Aires, Argentina

Research and Professional Experience:

Chemical kinetics and photochemistry. Optimization of drinking water and waste water treatment methods.

Professional Appointments:

Adjoint Researcher, CONICET-Argentina. Adjoint Professor, UNLP-Argentina.

Publications during last three years

- Evaluación del desempeño de un sistema continuo para el abatimiento de Arsénico empleando la técnica ZVI. Juan M. Triszcz*, Luciana Chippano, Enzo Pacho y Fernando S. García Einschlag ISA, AIDIS Argentina, Nº 104, 2009, 52-58 (ISSN: 0328-2937)
- Effect of solar-like radiation on the autocatalytic degradation of nitrobenzene using Fe^{+3} and hydrogen peroxide. Luciano Carlos, Debora Fabbri, Alberto L. Capparelli, Alessandra Bianco Prevot, Edmondo Pramauro and Fernando García Einschlag*. Journal of Photochemistry and Photobiology A. 2009, 201, 32-38 (ISSN: 1010-6030)
- Effect of operating conditions on iron corrosion rates in zero-valent iron systems for arsenic removal. Juan M. Triszcz, Andrés Porta and Fernando García Einschlag*. Chemical Engineering Journal, 2009, 150, 431-439 (ISSN: 1385-8947)
- Kinetics of nitrobenzene and 4-nitrophenol degradation by UV irradiation in the presence of nitrate and nitrite ions Fernando García Einschlag*, Juan Ignacio Felice and Juan Manuel Triszcz. Photochem. Photobiol. Sci., 2009, 8, 953-960. (ISSN: 1474-905X, DOI: 10.1039/B901835B)
- On the mechanism of the photoinduced reduction of and adduct of ferricytochrome C with a poly (4-vinylpyridine) polymer containing -ReI (CO)3(3,4,7,8-tetramethyl-1,10-phenanthroline) pendants. Larisa L. B. Bracco, Fernando S. García Einschlag, Ezequiel Wolcan* and Guillermo J. Ferraudi. Journal of Photochemistry and Photobiology A 2009, 208, 50-58 (ISSN: 1010-6030)
- Horseradish peroxidase and hematin as biocatalysts for alizarin degradation using hydrogen peroxide Silvina Pirillo*, Fernando Sebastián García Einschlag, Elsa H. Rueda, María Luján Ferreira *Ind. Eng. Chem. Res.* 2010, *49,* 6745–6752 (ISSN: 0888-5885), doi:10.1021/ie901528y
- Eriochrome Blue Black R and Fluorescein degradation by hydrogen peroxide oxidation with horseradish peroxidase and hematin as biocatalysts. Silvina Pirillo*, Fernando S. García Einschlag, María Luján Ferreira and Elsa H. Rueda. Journal of Molecular Catalysis B: Enzymatic *66 (2010) 63–71* (ISSN: 1381-1177), doi:10.1016/j.molcatb.2010.03.003
- Nitration of nitrobenzene in fenton's processes. Luciano Carlos, Daniela Nichela, Juan M. Triszcz, Juan I. Felice and Fernando S. García Einschlag* Chemosphere 2010, 80, 340-345 (ISSN: 0045-6535)
- Degradation kinetics of hydroxy- and hydroxynitro-derivatives of benzoic acid by fenton-like and photo-fenton techniques: a comparative study. Daniela Nichela, Ménana Haddou, Florence Benoit-Marquié, Marie-Thérèse Maurette, Esther Oliveros, Fernando S. García Einschlag* Applied Catalysis B: Environmental 98 (2010) 171–179 (ISSN: 0926-3373) doi:10.1016/j.apcatb.2010.05.026
- On the mechanism of formation and spectral properties of radical anions generated by the reduction of - [ReI(CO)3(5-Nitro-1,10-phenanthroline)]+and -[ReI(CO)3(3,4,7,8-

tetramethyl-1,10-phenanthroline)]+ pendants in poly-4-vinylpyridine polymers Larisa L.B. Bracco, Reynaldo O. Lezna, Jackeline Muñoz-Zuñiga, Gustavo T. Ruiz, Mario R. Féliz, Guillermo J. Ferraudi, Fernando S. García Einschlag*, Ezequiel Wolcan*. Inorganica Chimica Acta 370 (2011) 482-491 (ISSN: 0020-1693). doi:10.1016/j.ica.2011.02.039

Books & Book Chapters

- Procesos Químicos y Fotoquímicos ambientalmente benignos para la degradación de contaminantes en efluentes. *Chapter Title:* Métodos Fotoquímicos y Térmicos. (Capítulo 7) by Fernando S. García Einschlag, Luciano Carlos, Verónica Mora, Ana M. Amat, Rosa Vercher. Marfil, España, ISBN: 978-84-268-1485-2 (2009).
- Waste Water / Treatment and Reutilization. *Chapter Title:* Degradation of nitroaromatic compounds by homogeneous AOPs by Fernando Sebastián García Einschlag, Luciano Carlos and Daniela Alejandra Nichela. InTech. Open Access Publisher (2011). *Link:* www.intechopen.com/articles/show/title/degradation-of-nitroaromatic-compounds-by-homogeneous-aops *ISBN:* 978-953-307-249-4
- Waste Water / Treatment and Reutilization *Book Editor:* Fernando Sebastián García Einschlag. InTech. Open Access Publisher. (2011) *Link:* www.intechweb.org/books /show/title/waste-water-treatment-and-reutilization. *ISBN:* 978-953-307-249-4
- Waste Water / Evaluation and Management. *Book Editor:* Fernando Sebastián García Einschlag. InTech. Open Access Publisher. (2011) *Link:* www.intechweb.org/books /show/title/waste-water-evaluation-and-management *ISBN:* 978-953-307-233-3

In: Chemistry Researcher Biographical Sketches
Editors: E. S. Huseynov and V. Babayev
ISBN: 978-1-62100-771-5

Chapter 20

BIOGRAPHICAL SKETCH

Name:	Title:	Date of Birth:
Sami Fattouch	National Institute of Applied Sciences & Technology (INSAT)	January 11th 1970,

EDUCATION:

Institution and Location	Degree	Year Conferred	Scientific Field
(Ability to Direct Research). February 2010, INSAT, Tunis.	HDR		
Biochemistry and Molecular Biology, July 1999. Faculty of Sciences, Tunis.	Ph.D		
Genetic and Molecular Biology, December 1995. Faculty of Sciences, Tunis.	M.S.		
Biology, June 1993. Faculty of Sciences, Tunis, Tunisia.	B.S.		

CONTACT POINTS:

National Institute of Applied Sciences & Technology (INSAT), 1080 Tunis, Tunisia.
Phone: +21671703 829 (Office) / +21697558833 (Cellular); Fax: +216-71-704 329. E-mail: sami.fattouch@insat.rnu.tn

PROFESSIONAL EXPERIENCE:

2001-2010: Assistant Professor, Biochemistry (Option: Food Biochemistry, Molecular Biology and Bioinformatics); Department of Chemical and Biological Engineering; National Institute of Applied Science and Technology (INSAT). Tunis. Tunisia.

1999-2001: Assistant, Microbiology, Biochemistry and Genetics; Faculty of Sciences of Bizerte. Tunisia.

1996-1999: Instructor, Biochemistry; Faculty of Sciences of Tunis. Tunisia.

2008-2011: Member of the Scientific and Pedagogic Council of National Institute of Applied Science and Technology (INSAT). Tunis. Tunisia.

2007-2009: Head of the Department of Biological and Chemical Engineering at the Polytechnique of the Private University of Tunis (ULT). Tunisia.

Professional Appointments:

Since 24 January 2011) Senior Professor, Biochemistry (Option: Food Biochemistry, Molecular Biology and Bioinformatics); Department of Chemical and Biological Engineering; National Institute of Applied Science and Technology (INSAT). Tunis. Tunisia.

Publications during last three years

1- Dhaouadi, K., Raboudi, F., Estevan, C., Barrajon, E., Vilanova, E., Hamdaoui, M. and Fattouch, S. 2011. Cell Viability Effects and Antioxidant and Antimicrobial Activities of Tunisian Date Syrup (Rub El Tamer) Polyphenolic Extracts. *Journal of agricultural and food chemistry,* 59: 402-406.

2- Najjaa, H., Zerria, K., Fattouch, S., Ammar, E. and Neffati, M. 2011. Antioxidant and antimicrobial activities of *Allium roseum*. "lazoul", a wild edible endemic species in North Africa. *International Journal of Food Properties* (14): 371–380.

3- Belguith, H., Fattouch, S., Jridi, T. and Ben Hamida, J. 2009. Immunopurification of a rape (Brassica napus) seedling lipase. *African Journal of Biochemistry Research*, 3 (11): 356-365.

4- Fattouch, S., Sadok, S., Raboudi-Fattouch, F. and Ben Slama, M. 2008. Damage inhibition during refrigerated storage of mackerel (*Scomber scombrus*) fillets by a presoaking in quince (*Cydonia oblonga*) polyphenolic extract. *International Journal of Food Science and Technology,* 43: 2056–2064.

5- Ellouze, O., Fattouch, S., Mestiri, F., Aniba, M.R. and Marzouki, M.N. 2008. Optimization of extracellular xylanase production by *Sclerotinia sclerotiorum* S2 using factorial design. *Indian Journal of Biochemistry and Biophysics*, 45: 404-409.

6- Fattouch, S., Caboni, P., Coroneo, V., Tuberoso, C.I.G., Angioni, A., Dessi, S., Marzouki N. and Cabras, P. 2008. Comparative Analysis of Polyphenolic Profiles and Antioxidant and Antimicrobial Activities of Tunisian Pome Fruit Pulp and Peel Aqueous Acetone Extracts. *Journal of agricultural and food chemistry,* 56: 1084-1090.

7- Fattouch, S., Caboni, P., Coroneo, V., Tuberoso, C.I.G., Angioni, A., Dessi, S., Marzouki N. and Cabras, P. 2007. Antimicrobial Activity of Tunisian Quince (*Cydonia oblonga* Miller) Pulp and Peel Polyphenolic Extracts. *Journal of agricultural and food chemistry,* 55 (3): 963-969.

8- Fattouch, S., Acheche, H., M'Hirsi, S., Mellouli, L., Bejar, S., Marrakchi, M. and Marzouki, N. 2005. RT-PCR-RFLP for genetic diversity analysis of Tunisian Grapevine fanleaf virus isolates in their natural host plants. *Journal of Virological Methods,* 127(2):126-32.

9- Fattouch, S., H. Acheche, S. M'Hirsi, M. Marrakchi and N. Marzouki. 2005. Detection and characterization of two strains of Grapevine fanleaf nepovirus in Tunisia. *European Plant Pathology Office Bulletin,* 35 (2): 265-270.

10- M'Hirsi, S., Acheche, H., Fattouch, S., Boccardo, G., Marrakchi, M. and Marzouki, N. 2004. First report of phytoplasmas in the aster yellows group infecting grapevine in Tunisia. *Plant Pathology,* 53 (4) : 521-521

11- Smaali, M.I., Gargouri M., Limam, F., Fattouch, S., Maugard, T., Legoy, M.D. and Marzouki, N. 2003. Production, Purification, and Biochemical Characterization of Two β-Glucosidases From *Sclerotinia sclerotiorum. Applied Biochemistry and Biotechnology,* 111 (1): 29-40

12- Laarif, A., Fattouch, S., Essid, W., Marzouki, N., Ben Salah, H. and Ben Hamouda, M.H. 2003. Epidemiological survey of the *Phthorimaea operculell*a granulosis virus in Tunisia. *European Plant Pathology Office Bulletin,* 33 (2): 335-338

13- Saidaine, S., Fattouch, S., Hajlaoui, M.R., Marzouki, N. and Limam, F. 2002. Caractérisation biochimique d'enzymes hydrolytiques extracellulaires à partir du champignon phytopatogène *Sclerotinia sclerotiorum. Biologia (Tunisia),* 1, 21-27.

14- Fattouch, S., M'Hirsi, S., Acheche, H., Marrakchi, M. and Marzouki, N. 2001. RNA Oligoprobe Capture RT-PCR, a Sensitive Method for the Detection of Grapevine Fanleaf Virus in Tunisian Grapevines. *Plant Molecular Biology Reporter,* 19 *235-244.*

15- M'Hirsi S., Fattouch, S., Acheche, H., Marrakchi, M. and Marzouki, N. (2001). Detection of Grapevine A Vitivirus in Tunisian grapevines. *European Plant Pathology Office Bulletin,* 31 (4): 509-514

16- Belguith, H., Khodjet El Khil, H., Fattouch, S., Jridi, T. and Ben Hamida, J. 2001. Contribution of blotting techniques to the study of rapeseeds (*Brassica napus* L.) lipases. *Electrophoresis,* 22(1):18-22

17- Ferchichi, M., Limam, F., Fattouch, S., Marzouki, N. and Manai, M. 2000. Caractérisation et purification par HPLC d'une bactériocine produite par un lactocoque isolé de dérivés laitiers artisanaux. *Bulletin de la Société des Sciences Naturelles,* 29: 15-21.

18- Acheche, H., Fattouch, S., M'Hirsi, S., Marzouki, N. and Marrakchi, M. 1999. Use of optimised PCR methods for the detection of GLRAV3: a closterovirus associated with grapevine leafroll in Tunisian grapevine plants. *Plant Molecular Biology Reporter,* 17 (1): 31-42.

19- Acheche, H., Fattouch, S., M'Hirsi, S., Marzouki, N. et Marrakchi, M. (1998). Les viroses de la vigne en Tunisie (Grapevine viruses in Tunisia). *Archives de l'Institut Pasteur de Tunis, 75*: 219-226.

Project Coordinator/Member

- Coordinator of the Tunisian-Portugal project "Bioactive polyphenols from Pomegranate fruit: purification and identification of the main compounds; Use of the Yeast cell model to evaluate their protective properties against oxidative and heat stress"; in collaboration with the Faculdade de Farmacia, Universidade de Lisboa, Lisboa, Portugal (2009-2012).
- Coordinator of the Tunisian-Moroccan project 08/TM/38 "Implication of the Polyphenol Oxydase (PPO) in the Bioreactivity and the Stability of the Pesticides in the Food Matrices: Biochemical and Modeling study"; in collaboration with the Agronomic and Veterinary institute (IAV) Hassan II, Rabat, Morocco (2008-2011).
- Coordinator of the Tunisian-Egyptian project "Biotechnological Pesticide Transformation and Removal in soil, water and food: Biochemical and Molecular screening of pesticide degrading enzymes" in collaboration with the Genetic engineering and Biotechnology Institute, New Borg el Arab City, AlexAndria, Egypt (2007-2009)
- Coordinator of the Tunisian-Spanish project A/8823/07 "Cloning and Surexpression of Quince and Loquat Polyphenol Oxidases (PPO): Application in Food Quality Control and Development of Bbiosensor tools"; in collaboration with the Universidad d´Alacante, ALICANTE, Spain (2007-2008).
- Coordinator of the Tunisian-Spanish project A/9407/07 "Biotechnological Decontamination of Pesticide Residues: Enzymes Application and Biofilm Conception in Water Detoxification"; in collaboration with the Instituto de Bioingeniería, Universidad Miguel Hernández, Elche, Spain (2007-2008).
- Coordinator of the Tunisian-Moroccan project 06/TM/22 "Bioremediation of polluted water with pesticides residues from agricultural practices: chemical and biochemical detection of pesticide residues and biofilm confection"; in collaboration with the Faculté des Sciences d'Agadir, Agadir, Morocco (2006-2009).
- Member of the Tunisian-Spanish project A/2899/05 "Agro-food Application of Polyphenols and Polyphenol Oxidases (PPO)" in collaboration with the Universidad De Valencia, Valencia, Spain (2005-2006-2007).

Associative Scientific Activities and Honors

- Member of the Scientific committee of the "Conference on the bioactives substances", organized by BiotechPole / INRAP Sidi Thabet, Ariana, Tunisia; 3rd - 4th December, 2009 Hammamet.
- Co-organizer of the "6th Mediterranean Group of Pesticide Research (MGPR) Symposium MGPR", Cairo, Egypt, 27th - 30th October 2009.
- Member of the Scientific committee of the "Third International Medicinal and Aromatic Seedlings Symposium"; SIPAM - Jerba 26th - 28th March 2009.
- Co-organizer of the "Annual conference off the Mediterranean Group of Pesticide Residue (MGPR); Chemicals and their Residues in Food: New Scenarios of the Modern

Sustainable Agricultural Production 5th MGPR Symposium"; Piacenza (Italy), November 13th – 14th 2008.

- Co-organizer of the "5th Mediterranean Group of Pesticide Research (MGPR) Symposium", Agadir, Morocco, 21st - 24th November 2007.
- Member of the Office of the President and representing of Tunisia to the Mediterranean Group off Research Pesticide (MGPR): since 2004.
- Member of the MGPR (Mediterranean Group off Research Pesticide): since 2003.
- Member of the A.T.S.B (Tunisian Association of Biological Sciences): since 1997.
- Member of the A.T.B (Tunisian Association of Biotechnologies): since 2001.
- Member of the S.T.C.B (Tunisian Association of Biological Chemistry) :1994-1996.
- Association the Club of Young Sciences of Tunis of 1991-1995.

Recent Supervised Masters/Theses

Doctorate (PhD. equivalent)

- Dhaouadi Karima (since 2010). "Biochemical analysis of polyphenolic extracts of Tunisian date (*Phoenix dactylifera* L.), Barbary fig (*Opuntia ficus-indica*) and Carob (*Ceratonia siliqua*) syrups; evaluation of their cell viability effects and antioxidant and antimicrobial activities". INSAT, Tunis, Tunisia.
- Sioud Besma (since 2010). "Biochemical study of polyphenolic compounds from Tunisian pomegranate *Punica granatum* fruit derivatives : juice, jam and syrup" INSAT-ISA Sousse, Tunisia.
- Belkhir Manel (since 2010). "Biochemical analysis of Tunisia olive lives and azerole polyphenolic extracts: factorial design of mixtures; effect of irradiation process on phenolic content, antioxidant and antimicrobial activities". INSAT-FST, Tunis, Tunisia.
- Ben Mustapha Maha (2009-2011). "Quince latent polyphenol oxidase (PPO) cloning and expression: application to design a biosensor tool for phenolic compounds detection in food and environment matrices". INSAT-FSB, Tunisia.
- Ben Salem Issam (2008-2010). "Molecular tools to analyze microbial detoxification of heavy metals and pollutants, gamma irradiation effect n the process". INSAT-FSB-CNSTN, Tunisia.

Masters

- Ben Salem Issam (2006-2007). "Bioinformatic tools for the detection of pathogens sequences with PCR". INSAT-FSB-CNSTN, Tunisia.
- Ben Saad Rym (2006-2007). "Bioinformatic tools for the detection of food frauds using mitochondrial molecular markers". INSAT-FST, Tunis, Tunisia.
- Dhaouadi Karima (2006-2007). "The effect of heat treatment on the biological activities of green and black tea: biochemical analysis". INSAT – Health Institute, Tunis, Tunisia.

- Lajnef Imen (2008-2009). "Biochemical study of the polyphenolic extracts of "Henna" (*Lawsonia inermis*) a member of *Lythraceae* family". INSAT - Health Institute, Tunis, Tunisia.
- Nasralli Kamel (2008-2009). "Bioinformatic and molecular analysis of the *nucleopolyhedrovirus* (NPV) of *Spodoptera littoralis* in Tunisia". INSAT-ISA Sousse, Tunisia.
- Rebai Olfa (2008-2009). "Biochemical study of polyphenolic compounds from mulberry *Morus aba* & *Morus nigra* fruits in Tunisia". INSAT-FST, Tunis, Tunisia.
- Belkhir Manel (2008-2009). "Biochemical study of polyphenolic compounds from *Crataegus azarolus* a Rosacea member in Tunisia". INSAT-FST, Tunis, Tunisia.
- Sioud Besma (2008-2009). "Biochemical study of polyphenolic compounds from pomegranate *Punica granatum* fruit in Tunisia INSAT-ISA Sousse, Tunisia.
- Chourabi Khira (2008-2009). "Biochemical characterization of prophenol oxidase in the *Phtorimaea operculella (Zeller)* insect in Tunisia". INSAT-ISA Sousse, Tunisia.

In: Chemistry Researcher Biographical Sketches
Editors: E. S. Huseynov and V. Babayev
ISBN: 978-1-62100-771-5

Chapter 21

BIOGRAPHICAL SKETCH

Name:	Title:	Date of Birth:
Kikku Fukushima	Politecnico di Torino sede di Alessandria, Viale Teresa Michel 5, 15121 Alessandria, Italy	09/16/1981

EDUCATION:

Institution and Location	Degree	Year Conferred	Scientific Field
Post-Doctor on Biomaterials Engineering for Medical Applications (National Taiwan University of Science and Technology -NTUST-, Taiwan);			
Doctor on Materials Nanotechnology (University of Perugia, Italy);			
Materials Engineer and Master in Polymer Materials (Simon Bolivar University, Venezuela).			

CONTACT POINTS:

Da-An Road, Da-An District, No. 78, Taipei city 106, Taiwan (R.O.C).
106 台北市大安區大安路二段78號

AWARDS AND HONORS:

- Biography included in "Who's Who in the World 2011" (28th Edition). USA. (www.marquiswhoswho.com)

- Author of SciTopics (www.scitopics.com)

RESEARCH AND PROFESSIONAL EXPERIENCE:

During the last 6 years the main field of my research has been focused on the study of polymers, biopolymers and their nanocomposites for different industrial applications; such as packaging, food packaging, commodity and transport applications. This field includes: the study of mechanical, thermal, physical, chemical and morphological properties of polymers and their nanocomposites, especially of those systems concerning to biodegradable polymers (PLA, PCL, PBAT); new techniques for the preparation of polymer based nanocomposites with improved properties; study of thermal and mechanical degradation of polymers based materials; evaluation of biodegradation process of polymers in different natural media (soil, compost, microbial degradation); determination of the effect of nanoparticles on polymer (bio)degradation mechanism.

More recently, my field of research has been expanded to the study of biopolymers and their (nano)composites for possible biomedical applications, including: innovative biopolymer based systems for bone fracture treatments and for tissue engineering.

Professional Appointments:

- USB Polymer Group Researcher. Simón Bolívar University, Caracas - Venezuela. Duration: April-Oct, 2005. Laboratory Chief: Dr. Alejandro Müller.
- Research Assistant, Surfaces Engineering Center (CIS), Caracas – Venezuela. Duration: 2005-2006. Laboratory Chairman: Prof. Emeritus Joaquín Lira-Olivares.
- Research engineer, working with Prof. Jose Luis Feijoo - Simón Bolívar University (USB), Caracas – Venezuela. Duration: October 2005 – February 2006
- Researcher of the Politecnico di Torino, branch of Alessandria, -Italy, under the supervision of Prof. Giovanni Camino in different research projects about preparation, characterization and degradation of different bio-polymers and their nanocomposites. Duration: February 2006 - December 2009.
- Post-doctor researcher on Biomaterials Engineering for Medical Applications in National Taiwan University of Science and Technology (NTUST) - Taiwan. Duration: August 2010 – Actual position.

Publications during last three years

- “Nanocomposites of PLA and PCL based on montmorillonites and sepiolite”; K. Fukushima, D. Tabuani, G. Camino. *Materials Science and Engineering C,* 29, 1433 - 1441, (2009).
- “Biodegradation of Poly (lactic acid) and its nanocomposites”; K. Fukushima, C. Abbate, D. Tabuani, M. Gennari, G. Camino. *Polymer degradation and Stability*, 94, 1646-1655, (2009).

- "Biodegradation trend of Poly(ε-caprolactone) and nanocomposites"; K. Fukushima, C. Abbate, D. Tabuani, M. Gennari, P. Rizzarelli, G. Camino, *Materials Science and Engineering C,*30, 566-574, (2010).
- "Effect of expanded graphite / layered silicate clay on thermal, mechanical and fire retardant properties of Poly (lactic acid)"; Kikku Fukushima, Marius Murariu, Giovanni Camino, Philippe Dubois, *Polymer Degradation and Stability,* 95, 1063-1076, (2010).
- "Effect of sepiolite on the biodegradation trend of PLA and PCL"; Kikku Fukushima, Daniela Tabuani, Cristina Abbate, Maria Arena, Loredana Ferreri, *Polymer Degradation and Stability*, 95, 2049-2056, (2010).
- "Polylactic Acid and Polylactic Acid-Based Nanocomposite Photooxidation"; Sergio Bocchini, Kikku Fukushima, Alessandro Di Blasio, Alberto Fina, Alberto Frache, Francesco Geobaldo, *Biomacromolecules,* 11, 2919-2926, (2010).
- "Preparation, characterization and biodegradation of biopolymer nanocomposites based on fumed silica"; K. Fukushima, D. Tabuani, C. Abbate, M. Arena, P. Rizzarelli, *European Polymer Journal*, 47, 139-152, (2011).
- "Degradation of poly(lactic acid) and nanocomposites by *Bacillus licheniformis*"; Maria Arena, Cristina Abbate, Kikku Fukushima, Mara Gennari, accepted in *Environmental Science and Pollution Research*, DOI: 10.1007/s11356-011-0443-2, (2011)
- Invited book chapter "Biopolymer nanocomposites"; Kikku Fukushima, Daniela Tabuani, Cristina Abbate, in handbook titled *BIODEGRADABLE POLYMERS: Research and Applications,* by the Royal Society of Chemistry (RSC), expected data of publication: April – May 2011.

In: Chemistry Researcher Biographical Sketches
Editors: E. S. Huseynov and V. Babayev
ISBN: 978-1-62100-771-5

Chapter 22

BIOGRAPHICAL SKETCH

Name:	Title:	Date of Birth:
Alexander Y. Galashev	The main scientific employee of Institute of Industrial Ecology of the Ural Branch of the Russian Academy of Sciences	02 Sept. 1951

EDUCATION:

Institution and Location	Degree	Year Conferred	Scientific Field
Institute of Industrial Ecology, Ural Branch, Russian Academy of Sciences	PhD		

CONTACT POINTS:

620990 Yekaterinburg, Russia
Sofia Kovalevskaya Str. 20-a,

RESEARCH AND PROFESSIONAL EXPERIENCE:

Molecular dynamics simulation of water systems during 10 years and other systems during 35 years.

AWARDS AND HONORS:

2009 - The premium of the Russian Academy of Sciences for the best publication

Professional Appointments:

All positions which exist in the Russian Academy of Sciences from the lowest up to the highest.

Publications during last three years

1. Galashev A.Y., Rakhmanova O.R. Computer study of acetylene and ethane adsorption by disperse water medium 1/ Phys. Lett. A. 2008.V. 372. P. 3694-3698.
2. Galashev A.Y. Greenhouse effect of clusterization of CO2 and CH4 with atmospheric moisture // Environ Chem. Lett. 2009. DOI: 10.1007/s10311-009-0243-9.
3. Galashev A. Computer Study of Cluster Mechanism of Anti-greenhouse Effect 1/ Int.J.Env.Science and Engineering. 2009. V.1. N 1. pp. 31-38.
4. Galashev A.E. The spectral characteristics of chlorine-containing water clusters in the present of ozone and oxygen. // Russian Journal of Physical Chemistry A. 2009. V. 83. No 13. P. 55-60.
5. Galashev A.Y. Simulation of hydrocarbon molecules' absorption by ultra disperse water medium 1/ Molecular Simulation. 2010. V. 36, No. 4, P. 273-282.
6. Galashev A.Y. Simulation of medium-sized silicon clusters stabilized by hydrogen at high temperatures // International Journal of Mathematics, Game Theory and Algebra. 2009. V. 19. No 3. P. 193-211.
7. Galashev A.Y. The influence of atmospheric water clusters and halogen ions on ozone depletion // International Journal of Energy, Environment and Economics. 2010. V. 19. No 6. P. 1-21.

In: Chemistry Researcher Biographical Sketches
Editors: E. S. Huseynov and V. Babayev
ISBN: 978-1-62100-771-5

Chapter 23

BIOGRAPHICAL SKETCH

Name:	Title:	Date of Birth:
Julie Garden-Robinson		November 16, 1962

EDUCATION:

Institution and Location	Degree	Year Conferred	Scientific Field
North Dakota State University	Ph.D., Cereal Science and Food Technology	1994.	
North Dakota State University	M. S., Food/Nutrition with a minor in Biochemistry	1989.	
North Dakota State University	B.S. in Food and Nutrition/Dietetics with a minor in Music	1986.	

CONTACT POINTS:

North Dakota State University Extension Service
Health, Nutrition and Exercise Sciences - EML 351
NDSU Dept 7270 P.O. Box 6050
argo, ND 58108-6050

Awards and Honors:

- Distinguished Service Award – National Extension Association of Family and Consumer Sciences, October 2010.

- Extension Program of Excellence Award – NDSU Extension Service – "Barbecue Boot Camp" - 2009.
- National 1st Place Award Winner and Central Region 1st Place Winner – Marketing Award ("Eat Smart. Play Hard. Together). National Extension Association of Family and Consumer Sciences, October 2009.
- National 1st Place Award Winner and Central Region 1st Place Winner - Written Press Release ("Don't Fall Prey to the 5-second Rule). National Extension Association of Family and Consumer Sciences, October 2008.
- 2008 Exceptional Contributions to Research Award - College of Human Development and Education
- Extension Program of Excellence Award – NDSU Extension Service – "Eat Smart. Play Hard." educational program for children and families - 2007.

RESEARCH AND PROFESSIONAL EXPERIENCE:

Dr. Garden-Robinson has more than 17 years of professional experience as a faculty member and Extension Food and Nutrition Specialist at North Dakota State University. Her areas of research focus on nutrition, food safety and health especially for children and families. She serves on the North Dakota Nutrition Council, North Dakota Dietetic Association and March of Dimes Boards and writes a nutrition column that is used in several states.

Professional Appointments:

Associate Professor, Department of Health, Nutrition and Exercise Sciences and Adjunct Professor, School of Food Systems, North Dakota State University.

Publications during last three years

BOOK CHAPTERS

Lyonga, A.N., Eighmy, M. & Garden-Robinson, J. (2010). Culture and Food Safety Awareness: A Study of International College Students' Need for Food Safety Training in the U.S. In M.B. Walsch (Ed.), Food Supplies and Food Safety: Production, Conservation and Population Impact, Nova Publishers.

PEER REVIEWED PUBLICATIONS

Garden-Robinson, J. A. & Beauchamp, K. A. (2011, February). Folic acid every day: An educational intervention for women of childbearing age. *Family and Consumer Sciences Research Journal. 39*(3); pp. 246-255.

Stastny, S. & Garden-Robinson, J. (2011, March). Eating for Your Eyes. *Journal of Nutrition Education and Behavior. 43(2); pp. 137-139.*

Garden-Robinson, J., Eighmy, M., & Lyonga, A. (2010, December). Use of electronic group method in assessing food safety training needs and delivery methods among international college students in the U.S. Appetite. 55(3); pp.746-9.

Garden-Robinson, J. & Ussatis, R. (2010, March) On the Move to Better Health. *National 4-H Program of Distinction* Available from http://cyfernet.ces.ncsu.edu/pod/index.php

Lyonga, A.N., Eighmy, M.A., & Garden-Robinson, J. (2010, February) Examining the Prevalence of Self-Reported Foodborne Illnesses and Food Safety Risks among International College Students in the United States. *International Electronic Journal of Health Education.* 13:14.

Steinhaus, J., Brunt, A., Pankow, D., Garden-Robinson, J., & Terbizen, D. (2009, August) Who attends the Extension Family Nutrition Program in a rural state and what are the preferences of its diverse clientele? *Journal of Extension.* 47(4) Available from http://www.joe.org/joe/2009august/pdf/JOE_v47_4rb3.pdf

Abstract Publications

Stastny, S., Garden-Robinson, J., Youd, L. (Accepted for February 28, 2011). Effect of diabetic retinopathy education on knowledge and planned lifestyle activities. American Society of Nutrition: Advances and Controversies in Clinical Nutrition.

Garden-Robinson, J. A., Beauchamp, K. A. (2010, July). Folic acid every day intervention improved awareness, knowledge and self-reported behavior. *Journal of Nutrition Education and Behavior; 42* (4S), pp. S114-S115.

Garden-Robinson, J.A., Lyonga, A.N. & Eighmy, M. (2010, July). Examining the prevalence of self-reported foodborne illness and food safety risks among international college students in the United States. *Journal of Nutrition Education and Behavior;* 42(4S), pp. S114.

Osorno, J.M., Kelly, J.D., Brick, M., Urrea, C.A. Garden-Robinson, J., McClean, P. (2010, November). Strategies and challenges finding the next generation of plant breeders: The BeanCAP effort. ASA-CSSA-SSA Annual Meeting, Long Beach, CA.

Osorno, J.M., Kelly, J.D., Brick, M., Urrea, C.A. Garden-Robinson, J., McClean, P. (2010, August). Early recruitment of the next generation of plant breeders: The BeanCAP effort. 1st Annual Meeting of the National Association of Plant Breeders, Johnston, IA.

Garden-Robinson, J., Lyonga, A.N., & Eighmy, M. (2009, July) Use of Electronic Group Discussion Method to Assess Food-Related Educational Needs Among International College Students. *Journal of Nutrition Education and Behavior*, 41 (4S): S9.

Garden-Robinson, J., Schwarzwalter, K. & Gold, A. (2009, July) Food Handling Knowledge, Attitudes, and Behaviors Among Refugee New Americans. *Journal of Nutrition Education and Behavior*, 41 (4S): S39.

Stastny, S. & Garden-Robinson, J. (2009, March). Eating for your eye health. National Council on Aging, Las Vegas, NV.

Stastny, S. & Garden-Robinson, J. (2009, April). Eating for your eye health. United for Sight Conference, Yale Campus, New Haven, CT.

Garden-Robinson, J. (2008, September) Teens Serving Food Safely: A Curriculum to Improve Food Handling, Galaxy/National Extension Association of Family and Consumer Sciences, Indianapolis, IN.

Garden-Robinson, J. (2008, September) Banking on Strong Bones: An Educational Program to Improve Calcium Intake, Galaxy/National Extension Association of Family and Consumer Sciences, Indianapolis, IN.

Garden-Robinson, J. Almashari, H., Brunt, A.R., & Carlson, T. (2008, July) Factors impeding the implementation of safe food handling practices among New American populations. [Abstract] *Journal of Nutrition Education and Behavior,* 40(4S), pp. S65-S66.

Brunt, A.R., Steinhaus, J.L., & Garden-Robinson, J. (2008, July). Food stamp participant's barriers to participation and education preferences. *Journal of Nutrition Education and Behavior,* 40 (4S), pp. S65-S66.

Garden-Robinson, J. (2008, July) "Eat Smart. Play Hard." Using Role Models to Teach Healthy Behavior. *Journal of Nutrition Education and Behavior,* 40(4S), pp. S32.

In: Chemistry Researcher Biographical Sketches
Editors: E. S. Huseynov and V. Babayev
ISBN: 978-1-62100-771-5

Chapter 24

BIOGRAPHICAL SKETCH

Name:	Title:	Date of Birth:
Penka Dimitrova Gatseva	MD, PhD – Associate Professor	25 May 1956

EDUCATION:

Institution and Location	Degree	Year Conferred	Scientific Field
Medicine	MD		

CONTACT POINTS:

Medical University, 15A Vassil Aprilov Str., 4002 Plovdiv, Bulgaria

RESEARCH AND PROFESSIONAL EXPERIENCE:

Hygiene and Ecology, Environmental Medicine

Awards and Honors:

5 prizes for best reports on scientific forums

Professional Appointments:

Associate Professor, Department of Hygiene, Ecology and Epidemiology, Medical University, Plovdiv, Bulgaria

Publications during last three years

Articles 34;
Text- and handbooks for students 8;
Chapters in monographic books 1

In: Chemistry Researcher Biographical Sketches
Editors: E. S. Huseynov and V. Babayev
ISBN: 978-1-62100-771-5

Chapter 25

BIOGRAPHICAL SKETCH

Name:	Title:	Date of Birth:
Gennady Gerasimov		May 1, 1947

EDUCATION:

Institution and Location	Degree	Year Conferred	Scientific Field
Institute of Mechanics, Moscow State University			

CONTACT POINTS:

Michurinsky Avenue 1, 119192 Moscow, Russia

RESEARCH AND PROFESSIONAL EXPERIENCE:

Theoretical investigation of kinetic processes in physical and chemical gas dynamics
Awards and Honors:

Professional Appointments:

Leader Scientist, Doctor of Science

Publications during last three years

Gerasimov G. Modeling study of electron-beam VOCs decomposition inindustrial flue gases// Radiation Physics Research Progress/Ed. by A.N.Camilleri. New York: Nova Science Publishers, 2008. P. 247-286.

Nichipor G.V., Gerasimov G.Ya. Gas-phase radiation-chemical oxidationof naphthalene// High Energy Chemistry. 2008, 42(5), 335-341.

Gerasimov G. Ya., Pogosbekian Yu. M. Kinetics of thermal decomposition of organic binders in molds// J. Eng. Phys. Thermophys. 2009, 82(1), 92-97.

Gerasimov G. Ya. Comparative analysis of the formation mechanisms of polycyclic aromatic hydrocarbons and fullerenes in flames// J. Eng. Phys.Thermophys. 2009, 82(3), 432-441.

Gerasimov G. Mercury behavior in flue gases of coal-fired power plants: modeling study// Flue Gases: Research, Technology and Economics/ Ed. by T. B. Naylor. New York: Nova Science Publishers, 2009. P. 145-169.

Gerasimov G. Ya. Mechanisms of gas-phase degradation of polycyclic aromatic hydrocarbons by the action of ionizing radiation// High Energy Chemistry. 2010, 44(1), 1-5.

Gerasimov G. Ya. Radiation stability of carbon nanostructures//J. Eng. Phys. Thermophys. 2010, 83(2), 393-400.

Gerasimov G. Ya. Formation and conversion of carbon nanostructures under radiation// J. Eng. Phys. Thermophys. 2010, 83(4), 849-862.

Gerasimov G., Volkov E., Samuilov E. Thermodynamic and Kinetic Study of Oil Shale Processing// Engineering Physics and Mechanics: Analyses, Prediction and Applications, Eds. M. Sosa and J. Franco, N.Y., Nova Science Publishers, 2010. P. 473-492.

Gerasimov G. Radiation Induced Synthesis and Modification of Carbon Nanostructures// Engineering Physics and Mechanics: Analyses, Prediction and Applications, Eds. M. Sosa and J. Franco, N.Y., Nova Science Publishers, 2010. P. 493-508.

Volkov E., Gerasimov G., Ter-Oganesian G., Samuilov E. Investigation of flash pyrolysis of kukersite oil shale// Oil Shale. 2010, 27(4), 1-16.

In: Chemistry Researcher Biographical Sketches
Editors: E. S. Huseynov and V. Babayev
ISBN: 978-1-62100-771-5

Chapter 26

BIOGRAPHICAL SKETCH

Name:	Title:	Date of Birth:
Cecilia C. Guedes-Silva	Instituto de Pesquisas Energéticas e Nucleares IPEN-CNEN/SP.	April, 12, 1974.

EDUCATION:

Institution and Location	Degree	Year Conferred	Scientific Field
Bachelor's at Engenharia de Materiais from Universidade Presbiteriana Mackenzie (1997), Master's at Tecnologia Nuclear Aplicações from Instituto de Pesquisas Energéticas e Nucleares (2000) and Doctorate at Tecnologia Nuclear Materiais from Instituto de Pesquisas Energéticas e Nucleares (2005).			

CONTACT POINTS:

2242 Prof. Lineu Prestes Avenue, São Paulo, Brazil, ZIP Code 05508-000

RESEARCH AND PROFESSIONAL EXPERIENCE:

Processing and characterization of ceramic materials

Professional Appointments:

2010 - Current - Instituto de Pesquisas Energéticas e Nucleares - IPEN/USP – Researcher.
1998 – 2010 - Centro Tecnológico da Marinha Em São Paulo – CTMSP - Researcher.
2002 – 2009 - Academia Paulista Anchieta - UNIBAN (Universidade Bandeirante) – UNIBAN. – Teacher.

Publications during last three years

MARCHI, J. ; GUEDES e SILVA, CC ; SILVA, B.B. ; BRESSIANI, José Carlos ; BRESSIANI, Ana Helena. Influence of Additive System (Al2O3-RE2O3, RE = Y, La, Nd, Dy, Yb) on Microstructure and Mechanical Properties of Silicon Nitride-based Ceramics. Materials Research , v. 12, p. 145-150, 2009.

GUEDES e SILVA, CC ; BRESSIANI, José Carlos ; KÖNIG JR, Bruno . Estudo da Biocompatibilidade de Nitreto de Silício. In: KÖNIG Jr, B.. (Org.). Implantologia e Osteointegração Quanto à Pesquisa em Cerâmicas, Aplicações Clínicas e Outros Materiais. 1 ed. São Paulo: ROCA, 2010, p. 156-199.

In: Chemistry Researcher Biographical Sketches
Editors: E. S. Huseynov and V. Babayev
ISBN: 978-1-62100-771-5

Chapter 27

BIOGRAPHICAL SKETCH

Name:	Title:	Date of Birth:
Norma Guemes Vera	Instituto de Ciencias Agropecuarias UAEH	23-abril-1968

EDUCATION:

Institution and Location	Degree	Year Conferred	Scientific Field
	PhD		

CONTACT POINTS:

Gilberto Gímenez no 109-A Tulancingo Hidalgo Mexico

AWARDS AND HONORS:

Reconocimiento al Perfil Deseable- SEP 2003 a la fecha, Candidato al Sistema Nacional de Investigadores 2006-2009, Sistema Nacional Nivel 1 2010-2012.

RESEARCH AND PROFESSIONAL EXPERIENCE:

9 years with Investigator and Professor of UAEH

Professional Appointments:

Postgraduades and Investigator of Coordination of Instituto de Ciencias Agropecuarias UAEH

Publications during last three years

Effect of Temperature on Quality of Minimally Processed Litchi Fruit (*Litchi chinensis Sonn*). Publicado 2010. Journal of Food Quality.

Effect Of Kappa- and Lamda-Carrageenans as Fat-Replacers In Low-Fat Oaxaca Cheese. Alfonso Totosaus* and Norma Gímemes-Vera. 2008. International Journal of Food Properties

Utilización de los derivados de cereales y leguminosas en la elaboración de productos cárnicos. Norma Guemes Vera. NACAMEH en Ciencia y Tecnologia de la Carne. ISSBN-en trámite pag. 97-104.

Propiedades de textura de masa y pan dulce tipo "concha" fortificados con concentrados de suero de leche. Norma Guemes Vera, Alfonso Totosaus Sanchez, Juan Francisco Hernández Chavez, Sergio Soto Simental, Elia Nora Aquino Bolaños. Ciencia y Tecnología de Alimentos 2009.

Effective detoxificaction and decolouration of *L. mutabilis* seed derivatives, and effect derivatives on bread quality and acceptance. Noma Guemes Vera; Roberto Javier Peña Bautista; Cristian Jimenez Martinez and Gloria Davila Ortiz. Journal Agriculture of Food Science. 2008.:

Estudio del Efecto de la Adición de Sustitutos de Grasa (Maltodextrina) y Carbohidratos (Celulosa) en Pasteles de Chocolate. P. Neri Gómez, S.Soto Simental, M$^{a.}$ Isabel Reyes Santamaria, A. D. Hernandez Fuentes, E. N. Aquino Bolaños, N. Guemes Vera. 2do Encuentro Agroalimentario de la UAEH. Publicado 2006.ISBN 970-769-105-0.

In: Chemistry Researcher Biographical Sketches
Editors: E. S. Huseynov and V. Babayev

ISBN: 978-1-62100-771-5

Chapter 28

BIOGRAPHICAL SKETCH

Name:	Title:	Date of Birth:
Francisco Javier Guillén Gerada	University of Extremadura	27 August 1973

EDUCATION:

Institution and Location	Degree	Year Conferred	Scientific Field
	Ph.D. in Physics		

CONTACT POINTS:

LARUEX, Dpt. Applied Physics, Faculty of Veterinary Sciences, Avda. Universidad, s/n, 10003, Cáceres, Spain

AWARDS AND HONORS:

RESEARCH AND PROFESSIONAL EXPERIENCE:

Environmental Radioactivity

Professional Appointments:

Doctorate Lecturer

Publications during last three years

- J Guillén, A Baeza, N Mocanu, Analysis of the potential disadventages of chemical countermeasures against the transfer of radiocaesium to plants. Radioprotection 2009, 44(5), 543-546.
- A Baeza, J Guillén, MA Ontalba, A Rodríguez, FJ Ager, Radiological and multi-element analysis of the sediments from the Proserpina reservoir (Spain) dating from Roman times, J. Environ. Radioactiv. 2009, 100, 866-874.
- J Guillén, A Baeza, MA Ontalba, MP Miguez, ^{210}Pb and stable lead content in fungi: its transfer from soil. Sci. Total Environ. 2009, 407, 4320-4326.
- J Guillén, A Baeza , A Salas, Strontium., in Ed. D. Atwood: Radionuclides in the environment, 2010, John Wiley & Son, 79-96.
- A Baeza, I Vallejo, J Guillén, A Salas, JA Corbacho, Antlers of Cervus elaphus as biomonitors of ^{90}Sr in the environment. J. Environ. Radioactiv. 2011, 102, 311-315.
- A Baeza, J Miranda, J Guillén, JA Corbacho, A new approach to the analysis of the alpha spectra based on neural networks techniques, Nuclear Instruments and Methods in Physics Research A, 2011, in press.
- A Baeza, JA Corbacho, J Guillén, A Salas, Analysis of the different source terms of natural radionuclides in a river affected by NORM (Natural Occurring Radioactive Materials) activities, Chemosphere, 2011, in press.

In: Chemistry Researcher Biographical Sketches
Editors: E. S. Huseynov and V. Babayev
ISBN: 978-1-62100-771-5

Chapter 29

BIOGRAPHICAL SKETCH

Name:	Title:	Date of Birth:
Sergio Zarazua Guzman	Facultad de Ciencias Químicas, Universidad Autónoma de San Luis Potosi, Mexico	February 19th, 1980

EDUCATION:

2005-2009: PhD. Basic Biomedical Sciences, Medicine Faculty, UASLP. Mexico.
2003-2005: MSc. Basic Biomedical Sciences, Medicine Faculty, UASLP. Mexico.
1998-2002: Farmaceutical Chemistry, Chemistry Faculty, UASLP, Mexico.

CONTACT POINTS:

Av. Manuel Nava, 6. Zona Universitaria. CP. 78210, San Luis Potosí, S.L.P. Mexico.

RESEARCH AND PROFESSIONAL EXPERIENCE:

2010-2011: Post-doctoral associate, Molecular Biology Department, Instituto Potosino de Investigacion Cientifica y Tecnologica, Mexico.
2009-2010: Associate Professor, Universidad Cuauhtemoc Plantel San Luis Potosi, Mexico.

Professional Appointments:

Full Professor, Chemistry Faculty, UASLP. Mexico.

Publications during last three years

Valdez-Morales, E., Guerrero-Alba, R., Liñán-Rico, A., Espinosa-Luna, R., Miranda-Morales, M., Zarazúa-Guzmán, S., Barajas-López Carlos. P2X7 and P2X2 subunits contribute to the currents induced by ATP in guinea-pig intestinal myenteric neurons. Eur. J. Pharmacol. (Accepted: July 18, 2011).

Zarazúa, S., Bürger, S., Delgado, J.M., Jiménez-Capdeville, M.E., Schliebs, R. Arsenic affects expression and processing of amyloid precursor protein (APP) in primary neuronal cells overexpressing the Swedish mutation of human APP. J. Developmental Neuroscience. 2011, 29: 389-396.

Zarazúa, S., Ríos, R., Delgado, J.M., Santoyo, M.E., Ortiz-Pérez, D., Jiménez-Capdeville, M.E. Decreased arginine methylation and myelin alterations in arsenic exposed rats. Neurotoxicology. 2010, 31(1): 94-100.

Ríos, R., Zarazúa, S., Santoyo, M.E., Sepúlveda-Saavedra, J., Romero-Díaz, V., Jiménez, V., Pérez-Severiano, F., Vidal-Cantú, G., Delgado, J.M. and Jiménez-Capdeville M.E. Decreased nitric oxide markers and morphological changes in the brain of arsenic exposed rats. Toxicology, 2009, 261(1-2): 68-75.

In: Chemistry Researcher Biographical Sketches
Editors: E. S. Huseynov and V. Babayev
ISBN: 978-1-62100-771-5

Chapter 30

BIOGRAPHICAL SKETCH

Name:	Title:	Date of Birth:
Eduard Hanslík		

Publications during last three years

Hanslík, E., Ivanovová, D. Nuclear Power. Pavel Tsvetkov (ed.) Indie: Sciyo 2010. Impact of radionuclide discharges from Temelín Nuclear Power Plant on the Vltava River (Czech Republic), ISBN 978-953-307-110-7.

Diana Ivanovová, Eduard Hanslík Temporal and spatial changes in tritium concentration in the Vltava River basin affected by the operation of Temelín Nuclear Power Plant. Acta Universitatis Carolinae Environmentalica, 2010, 1-2, p. 17-31.

Ivanovová, D., Hanslík, E., Stierand P. The assessment of natural and artificial radionuclides in river sediments in the Czech Republic. Sediment Dynamics for a Changing Future, Vol. 337, 2010, p. 157-162.

Svetlik, I., Belanova, A., Vrskova, M., Hanslik, E., Ivanovova, D., Meresova, J., Tomaskova, L., Novakova, T. Volatility of 210Po in the gross alpha determination. Journal of Radioanalytical and Nuclear Chemistry, Vol. 286, 2010, 2, p. 547–551.

Hanslík, E.; Ivanovová, D.; Juranová, E.; Šimonek, P.; Jedináková-Křížová, V. Concentration of radionuclides in hydrosphere affected by Temelín nuclear power plant in Czech Republic. Journal of Environmental Radioactivity, Vol. 100, 2009, 7, p. 558-563.

Hanslík, E., Ivanovová, D., Juranová, E., Šimonek, P., Jedináková- Křížová V. Monitoring and assessment of radionuclide discharges from Temelín Nuclear Power Plant into the Vltava River (Czech Republic). Journal of Environmental Radioactivity, Vol. 100, 2009, 2, p. 131-138.

Hanslík, E., Ivanovová, D., Kluganostová, M. Balances of suspended matter and radionuclides in inflow and outflow waters of Orlík Reservoir, Vltava River (Czech Republic). Radioprotection, Vol. 44, 2009, 5, p. 321-326.

In: Chemistry Researcher Biographical Sketches
Editors: E. S. Huseynov and V. Babayev
ISBN: 978-1-62100-771-5

Chapter 31

BIOGRAPHICAL SKETCH

Name:	Title:	Date of Birth:
H. James Harmon	Department of Physics, Oklahoma State University, Stillwater, OK 73034 USA	9-14-1946

EDUCATION:

Institution and Location	Degree	Year Conferred	Scientific Field
BS, MS, PhD Purdue University			

CONTACT POINTS:

Department of Physics, Physics Sciences 145, Oklahoma State University, Stillwater, OK 73034 USA

RESEARCH AND PROFESSIONAL EXPERIENCE:

Sensors for detection of chemical and biological agents

Professional Appointments: Professor, Department of Physics, Oklahoma State University

Publications during last three years

2008 Early Events in TNT (2,4,6-Trinitrotoluene) Degradation by Porphyrins: Binding of TNT to Porphyrin by Hydrophobic and Hydrogen Bonds. Walid M. Hikal and H.J. Harmon. Journal of Hazardous Materials154, 826-831

2008 Role of Non-Linear Optics and Multiple Photon Absorbance in Enhancing Sensitivity of Enzyme-Based Chemical Agent Detectors. H. James Harmon. Methods 46, 18-24.

2009 Photocatalytic Self-Assembled Solid Porphyrin Microcrystals from Water-soluble Porphyrins: Synthesis, Characterization and Application. Walid. M. Hikal and H.J. Harmon. Polyhedron 28, 113-120.

2010 Crystal-growth Inhibition of Solid Porphyrin Micro-crystals by G4 PAMAM Dendrimer. Walid M. Hikal and H. James Harmon. J. Materials Science 46, 2273-2280.

2009 Advances in Porphyrin-based Photocatalytic 2,4,6-Trinitrotoluene Degradation. Jason Robinson and H. Ja mes Harmon. Chapter in "Energetic Materials: Chemistry, Hazards, and Environmental Aspects" Nova Science Publishers, Hauppage, NY. In Press

In: Chemistry Researcher Biographical Sketches
Editors: E. S. Huseynov and V. Babayev
ISBN: 978-1-62100-771-5

Chapter 32

BIOGRAPHICAL SKETCH

Name:	Title:	Date of Birth:
Masashi Hatanaka	Department of Green and Sustainable Chemistry, School of Engineering, Tokyo Denki University	1969/12/07

EDUCATION:

Institution and Location	Degree	Year Conferred	Scientific Field
Kyoto University (1989-1995), Tokyo Denki University (2004-2007)			

CONTACT POINTS:

Department of Green and Sustainable Chemistry, School of Engineering, Tokyo Denki University, 2-2 Kanda Nishiki-cho, Tokyo 101-8457

RESEARCH AND PROFESSIONAL EXPERIENCE:

Physical Chemistry, Quantum Chemistry

Professional Appointments:

Part-time lecturer (Tokyo Denki University)

Publications during last three years

M. Hatanaka, "Electronic states of graphene-based ferromagnets" in "Graphene/Book 1", in press.

In: Chemistry Researcher Biographical Sketches
Editors: E. S. Huseynov and V. Babayev
ISBN: 978-1-62100-771-5

Chapter 33

BIOGRAPHICAL SKETCH

Name:	Title:	Date of Birth:
Robert B. Heimann	Professor emeritus	Dec 31, 1938

RESEARCH AND PROFESSIONAL EXPERIENCE:

Am Stadtpark 2A, D-02826 Görlitz, Germany

In: Chemistry Researcher Biographical Sketches
Editors: E. S. Huseynov and V. Babayev
ISBN: 978-1-62100-771-5

Chapter 34

BIOGRAPHICAL SKETCH

Name:	Title:	Date of Birth:
Kazutaka Hirakawa	Shizuoka University	January 14, 1973

EDUCATION:

Institution and Location	Degree	Year Conferred	Scientific Field
Graduate School of Arts and Sciences, The University of Tokyo			

CONTACT POINTS:

Johoku 3-5-1, Naka-ku, Hamamatsu, Shizuoka 432-8561, Japan

AWARDS AND HONORS:

CSJ Presentation Award 2007 (2007, form The Chemical Society of Japan)
Award from Japanese Society for Photomedicne and Photobiology (2009)
IJRC Award (2010, from Innovation and Joint Research Center, Shizuoka University)

RESEARCH AND PROFESSIONAL EXPERIENCE:

2000-2004: Mie University School of Medicine,
2004-Faculty of Engineering, Shizuoka University

Professional Appointments:

Associate Professor

Publications during last three years

1. K. Hirakawa, T. Hirano, Y. Nishimura, T. Arai, Y. Nosaka, Control of Singlet Oxygen Generation Photosensitized by meso-Anthrylporphyrin through Interaction with DNA, *Photochem. Photobiol.*, 2011.(in press)
2. J. Matsumoto, T. Shinbara, S. Tanimura, T. Matsumoto, T. Shiragami, H. Yokoi, Y. Nosaka, S. Okazaki, K. Hirakawa, M. Yasuda, Water-soluble phosphrousporphyrins with high activity for visible light-assisted inactivation of *Saccharomyces cerevisiae*, *J. Photochem. Photobiol., A: Chem.*, 218, 178, 184, 2011.
3. K. Hirakawa and H. Segawa, Effects of hydrogen bonding interaction and solvent polarity on the competition between excitation energy and photo-induced electron transfer processes in hydroxy(1-pyrenebutoxy)phosphorus(V)porphyrin, *J. Photochem. Photobiol., A: Chem.*, 213, 73-79, 2010.
4. K. Hirakawa and H. Segawa, Tetra(*p*-*t*-butyl)thiacalix[4]arenetetrol Forms 4:1 Complex with Tetrakis(1-methyl-4-pyridinio)porphyrin, *Bull. Chem. Soc. Jpn.*, 83, 828-830, 2010.
5. K. Hirakawa and H. Segawa, Acid Dissociation of Axial Hydroxyl Group of Hydroxy(1-pyrenebutoxy)-phosphorus(V) Porphyrin Controls the Intramolecular Excitation Energy Transfer, *Photochem. Photobiol. Sci.*, 9, 704-709, 2010.
6. K. Hirakawa, K. Saito, and H.i Segawa, Anomalously Selective Quenching of S_2 Fluorescence from Upper Excited State of Zinc 5-(1(B-pyrenyl)-10,15,20-triphenylporphyrin Derivatives through Intramolecular Charge Transfer State I, *J. Phys. Chem. A*, 113, 8852-8856, 2009.
7. K. Hirakawa and S. Sano, Platinum Nanoparticle Catalyst Scavenges Hydrogen Peroxide Generated from Hydroquinone, *Bull. Chem. Soc. Jpn.*, 82, 1299-1303, 2009.
8. K. Hirakawa, Fluorometry of Singlet Oxygen Generated via a Photosensitized Reaction Using Folic Acid and Methotrexate, *Anal. Bioanal. Chem.*, 393, 999-1005, 2009.
9. K. Hirakawa, Y. Ebara, T. Hirano, and H. Segawa, Photo-damage of Human Serum Albumin by Porphyrin P(V) Complex through Electron Transfer and Singlet Oxygen Generation, *J. Jpn Soc. Laser Surgery and Medicine* 29, 372-378, 2009.
10. M. Yasuda, T. Nakahara, T. Matsumoto, T. Shiragami, J. Matsumoto, H. Yokoi, T. Hirano, and K. Hirakawa, Visible light-assisted Sterilization Activity of Water-soluble Antimonyporphyrin toward *Saccharomyces Cerevisiae*, *J. Photochem. Photobiol., A: Chem.*, 205, 210-214, 2009.
11. S. Tada-Oikawa, S. Oikawa, J. Hirayama, K. Hirakawa, and S. Kawanishi. DNA Damage and Apoptosis Induced by Photosensitization of 5,10,15,20-Tetrakis (*N*-methyl-4-pyridyl)-21*H*,23*H*-porphyrin *via* Singlet Oxygen Generation, *Photochem. Photobiol.*, 85, 1391-1399, 2009.

12. K. Hirakawa and T. Hirano. The Microenvironment of DNA Switches the Activity of Singlet Oxygen Generation Photosensitized by Berberine and Palmatine, *Photochem. Photobiol.*, 84, 202-208, 2008.

In: Chemistry Researcher Biographical Sketches
Editors: E. S. Huseynov and V. Babayev
ISBN: 978-1-62100-771-5

Chapter 35

BIOGRAPHICAL SKETCH

Name:	Title:	Date of Birth:
Tomohisa Hirobe	Active Radiation Protection Research Team	September 25, 1950

EDUCATION:

Institution and Location	Degree	Year Conferred	Scientific Field
Chiba University, Biology	Bachelor of Science,	1974	
Tohoku University, Biology	Master of Science,	1976	
Tohoku University, Developmental Biology	Doctor of Science	1979	

CONTACT POINTS:

Active Radiation Protection Research Team,
Radiation Risk Reduction Research Program,
National Institute of Radiological Sciences,
4-9-1 Anagawa, Inage-ku, Chiba 263-8555, Japan
Tel: +81-43-206-3253 or 3133
Fax: +81-43-206-4638
E-mail: thirobe@nirs.go.jp

RESEARCH AND PROFESSIONAL EXPERIENCE:

1977-1982 Assistant, Iwate University, Faculty of Education, Department of Biology
1982-1987 Lecturer, Iwate University, Faculty of Education, Department of Biology

1985-1986 Visiting Scientist (Sabbatical), Harvard School of Dental Medicine, Laboratory of Electron Microscopy

1987-2001 Senior Researcher, National Institute of Radiological Sciences, Division of Biology

1997-2011 Professor, Chiba University, Graduate School of Science

2001-2011 Team Leader, National Institute of Radiological Sciences, Radiation Effect Mechanisms Research Group

2011-present Active Radiation Protection Research Team, Radiation Risk Reduction Research Program, National Institute of Radiological Sciences,

Publications during last three years

1. 2003 Award, Ministry of Education, Culture, Sports, Science and Technology of Japan Publication Last 3 Years: 1.T. Hirobe, K. Ishizuka, S. Ogawa and H. Abe. Mitochondria are more numerous and smaller in pink-eyed dilution melanoblasts and melanocytes than in wild-type melanocytes in the neonatal mouse epidermis. Zool. Sci. 25(11):1057-1065(2008), November
2. T. Hirobe. Ferrous ferric chloride induces the differentiation of cultured mouse epidermal melanocytes additionally with herbal medicines. J. Health Sci. 55(1):86-94(2009), February
3. T. Hirobe. Ferrous ferric chloride stimulates the proliferation of human skin keratinocytes, melanocytes, and fibroblasts in culture. J. Health Sci. 55(3):447-455(2009), June
4. R. Furuya, Y. Yoshida, O. Moro, M. Tsunenaga, H. Aoki, J. Kishimoto, O Ifuku, and T. Hirobe.Immunohistochemical survey of the distribution of epidermal melanoblasts and melanocytes during the development of UVB-induced pigmented spots. J. Dermatol. Sci. 55(2): 99-107(2009), August
5. T. Hirobe. Ferrous ferric chloride stimulates the skin cell function and hair growth in mice. Biol. Pharm. Bull. 32(8):1347-1453(2009), August
6. F. Zakeri and T. Hirobe. A cytogenetic approach to the effects of low levels of ionizing radiations on occupationally exposed individuals. Eur. J. Radiol. 73(1): 191-195 (2010), January
7. T. Hirobe, T. Shinpo, K. Higuchi, T. Sano. Life cycle of human melanocytes is regulated by endothelin-1 and stem cell factor in synergy with cyclic AMP and basic fibroblast growth factor. J. Dermatol. Sci. 57(2):123-131 (2010), February
8. T. Hirobe, M. Kiuchi, K. Wakamatsu, S. Ito. Estrogen increases hair pigmentation in female recessive yellow mice. Zool. Sci. 27(6):470-476 (2010), June
9. F. Zakeri, T. Hirobe, K. Akbari Noghabi. Biological effects of low-dose ionizing radiation exposure on interventional cadiologists. Occup. Med. 60(6):464-469 (2010), September
10. T. Hirobe, S. Ito, K. Wakamatsu. The mouse pink-eyed dilution allele of the *P*-gene greatly inhibits eumelanin, but not pheomelanin synthesis as studied by chemical analysis. Pigment Cell Melanoma Res. 24(1):241-246 (2011) February

11. T. Hirobe, K. Eguchi-Kasai, K. Sugaya, M. Murakami. Effects of low-dose gamma-rays on the embryonic development of mouse melanoblasts and melanocytes in the epidermis and hair bulbs, Zoological Science, in press
12. T. Hirobe, K. Eguchi-Kasai, K. Sugaya, M. Murakami. Effects of low-dose heavy ions on the postnatal development of mice and the yield of white spots in the mid-ventrum and tail-tips, J. Radiat. Res., in press
13. T. Hirobe: How are proliferation and differentiation of mouse melanocytes regulated? Pigment Cell Melanoma Res., in press
14. T. Hirobe: Stimulation of the Proliferation and Differentiation of Skin Cells by Ferrous Ferric Chloride from a Distance. Biol. Pharm. Bull., in press.
15. T. Motohashi, K. Yamanaka, K. Chiba, K. Miyajima, H. Aoki, T. Hirobe, T. Kunisada: Neural crest cells retain their capability for multipotential differentiation even after lineage-restricted stages. Dev. Dynam., in press
16. T. Hirobe, C. Yoshihara, S. Takeuchi, K. Wakamatsu, S. Ito, H. Abe, Y. Kawa, Y. Soma: A novel deletion mutation of mouse ruby-eye 2 named *ru2d/Hps5ru2-d* inhibits melanocyte differentiation and its impaired differentiation is rescued by L-tyrosine. Zool. Sci., in press.

In: Chemistry Researcher Biographical Sketches
Editors: E. S. Huseynov and V. Babayev

ISBN: 978-1-62100-771-5

Chapter 36

BIOGRAPHICAL SKETCH

Name:	Title:	Date of Birth:
Gorazd Hribar	Department for biosynthesis and biotransformation, National Institute of Chemistry, Hajdrihova 19, Ljubljana, Slovenia	7th August 1980

EDUCATION:

Institution and Location	Degree	Year Conferred	Scientific Field
PhD in Biochemistry and Molecular Biology			

CONTACT POINTS:

National Institute of Chemistry, Hajdrihova 19, Ljubljana, Slovenia

RESEARCH AND PROFESSIONAL EXPERIENCE:

Researcher at National Institute of Chemistry

<u>Publications during last three years</u>

1. Hribar G, Smiloviæ V, Lenassi Zupan A, Gaberc-Porekar V. [beta]-lactamase reporter system for selecting high-producing yeast clones. BioTechniques. 2008, 44 (4), 477-483.

2. Bele M, Hribar G, Campelj S, Makovec D, Gaberc-Porekar V, Zorko M, Gaberscek M, Jamnik J, Venturini P. Zinc-decorated silica-coated magnetic nanoparticles for protein binding and controlled release. Journal of chromatography B: Analytical technologies in the biomedical and life sciences. 2008, 867 (1), 160-164.
3. Starèeviæ Š, Kocbek P, Hribar G, Lanišnik-Rižner T, Gobec S. Biochemical and biological evaluation of novel potent coumarin inhibitor of 17ß-HSD type 1. Chem.-biol. interact. 2011, doi: 10.1016/j.cbi.2011.01.002.
4. Hribar G, Žnidaršic A, Bele M, Maver U, Caserman S, Gaberšèek M, Gaberc-Porekar V. Zinc-phosphate nanoparticles with reversibly attached TNF- [alfa] analogs : an interesting concept for potential use in active immunotherapy. J. nanopart. res.. 2011, doi: 10.1007/s11051-010-0199-5.

In: Chemistry Researcher Biographical Sketches
Editors: E. S. Huseynov and V. Babayev
ISBN: 978-1-62100-771-5

Chapter 37

BIOGRAPHICAL SKETCH

Name:	Title:	Date of Birth:
Bouraoui Ilahi	PhD	30-AUG-1976

EDUCATION:

Institution and Location	Degree	Year Conferred	Scientific Field
University of Monastir			

CONTACT POINTS:

Laboratory of Micro-Optoelectronic and Nanostructures (LMON), Faculty of Sciences, Avenue de l'environnement, 5019 Monastir, Tunisia

RESEARCH AND PROFESSIONAL EXPERIENCE:

Professional affiliations

- Section Editor: Physical sciences, Akademeia (Sept. 2010- Aug. 2011)
- Editorial board member, Applied Physics Research (2010-2013)
- Technical Editor, Journal of Applied Sciences (2006-2009)

Technical Journals reviewer

- Nanotechnology
- New journal of Physics
- Journal of Physics D: Applied Physics
- Journal of Applied Physics

Research highlights

- Have successfully fabricated and optimized multi-layer electronically and/or mechanically coupled quantum dot heterostructures for optoelectronic application.
- Have successfully grown and demonstrated the possibility to extend the wavelength of InAs quantum dots on GaAs substrate to the telecom window.
- Carried out extensive post growth rapid thermal treatments to tune the QDs emission wavelength and quantum efficiency.
- Demonstrated the highest selectively tunable QDs emission energy by means of ion implantation and thermal annealing to use for broad band photodetector and supperluminescent diode.
- setting up of photomodulated reflectance experience and application for the characterization of micro and nanostructured materials and devices
- Conducted extensive MBE of GaAs-based strained heterostructures leading to the growth of state- of-the-art diluted quantum dot for single photon emitter and detector
- Study of the impact of the surrounding materiel's strain and composition on the morphology and emission wavelength of InAs quantum dots on GaAs substrate.
- Extensive investigation of core-shell InAs/InP nanowires on Si substrate for fibe optic telecom application

Professional Appointments:
Associate Professor

Publications during last three years

1. –«InP nanowires with InAs insertion grown by catalyst assisted molecular beam epitaxy on Si substrate» H. Khmissi, K. Naji, M. H. Haj Alouane, N. Chauvin, C. Bru-Chevallier, B. Ilahi, G. Patriarche and M. Gendry Journal of Applied Physics (submitted)
2. –« InP/InAs/InP core-shell nanowires emitting in telecommunication wavelength on Si substrate» M. H. Hadj Alouane, N. Chauvin, H. Khmissi, K. Naji , B. Ilahi, H. Maaref, G. Patriarche, M. Gendry, and C. Bru-Chevallier Nanotechnology(submitted)
3. –«InAs quantum dots on different Ga(In)As surrounding material investigated by photoreflectance and photoluminescence spectroscopy: Electronic energy levels and carrier's dynamic » M. H. Hadj Alouane, B. Ilahi, L. Sfaxi and H. Maaref, Journal of Nanoparticle Research (Accepted for publication)
4. –«Temperature dependent photoluminescence properties of InAs/InP quantum sticks subjected to low energy phosphorous ion implantation and subsequent annealing» M. H. Hadj Alouane, B. Ilahi, H. Maaref, B. Salem, V. Aimez, D. Morris and M. Gendry Journal of Nanoscience and Nanotechnology (accepted for publication)
5. –«Impact of ion-implantation-induced band gap engineering on the temperature dependent photoluminescence properties of InAs/InP quantum dashes» M. H. Hadj Alouane, B. Ilahi, H. Maaref, B. Salem, V. Aimez, D. Morris, A. Turala, P. Regreny and M. Gendry Journal of Applied Physics 108, 024317 (2010)

6. –«Persistence of In/Ga intermixing beyond the emission energy blueshift saturation of proton-implanted InAs/GaAs quantum dots» Z. Zaâboub, B. Ilahi,, B. Salem, V. Aimez, D. Morris L. Sfaxi, and H. Maaref, Journal of Applied Physics 107, 124306 (2010)
7. –« Intermixing of InAs/GaAs quantum dots by proton implantation and rapid thermal annealing» B. Ilahi, Z. Zaâboub, B. Salem, D. Morris, V. Aimez, L. Sfaxi, and H. Maaref Materials Science in Semiconductor Processing 12 (1-2), pp. 71-74 (2009)
8. –«Inhomogeneous broadening and alloy intermixing in low proton doses implanted InAs/GaAs self-assembled quantum dots» Z. Zaâboub, B. Ilahi, L. Sfaxi, H. Maaref, B. Salem, V. Aimez, and D. Morris Nanotechnology 19, 285715 (2008)
9. -«Optical investigation of phosphorous-ion-implantation induced InAs/GaAs quantum dots' intermixing» Z. Zaâboub, B. Ilahi, L. Sfaxi, H. Maaref, B. Salem, V. Aimez, and D. Morris Physics Letters A 372, 4714–4717 (2008)
10. -«Thermal-induced intermixing effects on the optical properties of long wavelength low density InAs/GaAs quantum dots» Z. Zaâboub, B. Ilahi, L. Sfaxi, H. Maaref, Material Sciences and Engineering C 28, 1002 (2008)

In: Chemistry Researcher Biographical Sketches
Editors: E. S. Huseynov and V. Babayev

ISBN: 978-1-62100-771-5

Chapter 38

BIOGRAPHICAL SKETCH

Name:	Title:	Date of Birth:
Piera Iuliani	PhD	07/05/1973

EDUCATION:

Institution and Location	Degree	Year Conferred	Scientific Field
Università degli Studi "G. D'Annunzio" Chieti-Pescara, Facoltà di Farmacia-Dipartimento di Scienze del Farmaco, Via dei Vestini, 66100 Chieti- Italy			

CONTACT POINTS:

Università degli Studi "G. D'Annunzio" Chieti-Pescara,
Via dei Vestini, 66100 Chieti- Italy

<u>Publications during last three years</u>

1. Development of a method for the determination of vardenafil in human plasma by high performance liquid chromatography with UV detection?
2. Carlucci G, Palumbo P, Iuliani P, Palumbo G, Biomedical Chromatography 2009, 23 (7), 759-763.
3. Investigation of the HPLC response of NSAIDs by fractional experimental design and multivariate regression analysis. Response optimization and new retention parameters? Iuliani P, Carlucci G, Marrone A, Journal of pharmaceutical and biomedical analysis 2010, 51 (1), 46-55, epub 8 agosto 2009.

4. Simultaneous determination of zofenopril and hydrochlorothiazide in tablets using derivative UV Spectrophotometry? Carlucci G, Di Federico L, Iuliani P, Analytical Letters 2010, 43, 2609-2617.
5. Simultaneous determination of benzydamine hydrochloride and five impurities in an oral collutory as a pharmaceutical formulation by high-performance liquid chromatography? Carlucci G, Iuliani P, Di Federico L, Journal of Chromatographic Science 2010, 48, 854-859.
6. HPLC-DAD method for the simultaneous determination of zofenopril and hydrochlorothiazide in oral pharmaceutical formulations? Carlucci G, Di Federico L, Iuliani P, Journal of Separation Science 2010, 33 (12), 1717-1722.
7. RP-HPLC method with fluorescence detection for amino acids D/L ratio determination in fossil bones as a marker of DNA preservation? Iuliani P, Di Federico L, Fontecchio G, Carlucci G, Journal of Separation Science 2010, 33 (16), 2411-2416.

In: Chemistry Researcher Biographical Sketches
Editors: E. S. Huseynov and V. Babayev

ISBN: 978-1-62100-771-5

Chapter 39

BIOGRAPHICAL SKETCH

Name:	Title:	Date of Birth:
Songsri Kaewsuwan	Department of Pharmacognosy and Pharmaceutical Botany, Faculty of Pharmaceutical Sciences, Prince of Songkla University, THAILAND	August 4th 1977

EDUCATION:

Institution and Location	Degree	Year Conferred	Scientific Field
Pharmaceutical Chemistry and Phytochemistry	Ph.D		
Pharmacognosy	M.Sc		
Pharm. (First Class Honors)	B.Sc		

CONTACT POINTS:

Department of Pharmacognosy and Pharmaceutical Botany, Faculty of Pharmaceutical Sciences, Prince of Songkla University, Hat Yai, Songkhla 90112, THAILAND

RESEARCH AND PROFESSIONAL EXPERIENCE:

Factors affecting polyunsaturated fatty acid (PUFA) production by lower plants and microorganisms

- Polyunsaturated fatty acid (PUFA) pathway: Gene cloning, expression and metabolic engineering
- Medicinal plant tissue culture and micropropagation

Professional Appointments:

Optimization of polyunsaturated fatty acid (PUFA) production by lower plants and microorganisms

- Production of novel PUFAs in lower plants

Publications during last three years

Kaewsuwan S, Bunyapraphatsara N, Cove DJ, Quatrano RS, Chodok P. High level production of adrenic acid in *Physcomitrella patens* using the algae *Pavlova* sp. Δ^5-elongase gene . Bioresour Technol 2010; 101(11): 4081-4088, 2010.

Chodok P, Kanjana-Opas A, Kaewsuwan S. The Plackett-Burman design for evaluating the production of polyunsaturated fatty acids by *Physcomitrella patens*. J Am Oil Chem Soc 2010; 87(5): 521-529.

Saelao S, Kanjana-Opas A, Kaewsuwan S. Optimization of biomass and arachidonic acid production by *Aureispira maritima* using response surface methodology. J Am Oil Chem Soc 2011; 88(5): 619-629.

In: Chemistry Researcher Biographical Sketches
Editors: E. S. Huseynov and V. Babayev

ISBN: 978-1-62100-771-5

Chapter 40

BIOGRAPHICAL SKETCH

Name:	Title:	Date of Birth:
C. Karunakaran	M.Sc., M.Phil., Ph.D	May 28th 1952

EDUCATION:

Institution and Location	Degree	Year Conferred	Scientific Field
Department of Chemistry, Annamalai University, Annamalainagar 608002, Tamilnadu, India			

CONTACT POINTS:

E-mail: karunakaranc@rediffmail.com

RESEARCH AND PROFESSIONAL EXPERIENCE:

36 years of teaching and research experience; guided 12 Ph.Ds and 26 M.Phils; completed many major research projects; authored more than 115 research articles and 3 books. Current research interest: Synthesis of nanomaterials for potential applications.

AWARDS AND HONORS:

Best Teacher Award 2009

Professional Appointments:

Professor and Head, Department of Chemistry, Annamalai University; Sectional Secretary - Physical Sciences, 94th Indian Science Congress, 2007.

Publications during last three years

1. 1.C. Karunakaran, V. Rajeswari, P. Gomathisankar, Combustion synthesis of ZnO and Ag-doped ZnO and their bactericidal and photocatalytic activities• Superlattices and Microstructures, doi: 10.1016/j.spmi.2011.06.005
2. C. Karunakaran, G. Abiramasundari, P. Gomathisankai, G. Manikandan, V. Anandi, Preparation and characterization of ZnO-TiO2 nanocomposite for photocatalytic disinfection of bacteria and detoxification of cyanide under visible light• , Materials Research Bulletin, doi: 10.1016/j.materresbull.2011.06.019
3. C. Karunakaran, P. Anilkumar, P. Gomathisankar, Photoproduction of iodine with nanoparticulate semiconductora and insulators• , Chemistry Central Journal, 2011, 5:31; doi: 10.1186/1752-153x-5-31.
4. C. Karunakaran, P. Gomathisankar, G. Manikandan, Solar photocatalytic detoxification of cyanide with bacterial disinfection by oxide ceramics• , Indian Journal of Chemical Technology, 2011, 18, 169-176.
5. C. Karunakaran, V. Rajeswari, P. Gomathisankar, Optical, electrical, photocatalytic, and bactericidal properties of microwave synthesized nanocrystalline Ag-ZnO and ZnO• , Solid State Sciences, 2011, 13, 923-928.
6. C. Karunakaran, V. Rajeswari, P. Gomathisankar, Enhanced photocatalytic and antibacterial activities of sol-gel synthesized ZnO and Ag-ZnO• , Materials Science in Semiconductor Processing, 2011, 14, 133-138.
7. C. Karunakaran, R. Dhanalakshmi, P. Gomathisankar, G. Manikandan, Photodegradation of carboxylic acids on Al2O3 and SiO2 nanoparticles• , Indian Journal of Chemistry A, 2011, 50, 163-170.
8. C. Karunakaran, A. Vijayabalan, G. Manikandan, P. Gomathisankar, Visible light photocatalytic disinfection of bacteria by Cd-TiO2• , Catalysis Communications, 2011, 12, 826-829.
9. C. Karunakaran, G. Abiramasundari, P. Gomathisankar, G. Manikandan, V. Anandi, NiO/TiO2 nanoparticles for photocatalytic disinfection of bacteria under visible light• , Journal of the American Ceramic Society, doi: 10.1111/j.1551-2916.2011.04403.x
10. C. Karunakaran, P. Gomathisankar, G. Manikandan, Solar photocatalytic detoxification of cyanide by different forms of TiO2• , Korean Journal of Chemical Engineering, 2011, 28, 1214-1220.
11. C. Karunakaran, V. Rajeswari, P. Gomathisankar, Photodeposited surface Ag on ZnO nanocrystals and the optical, electrical, photocatalytic, and bactericidal properties• , Synthesis and Reactivity in Inorganic, Metal-Organic, and Nano-Metal Chemistry, 2011, 41, 369-325.

12. C. Karunakaran, V. Rajeswari, P. Gomathisankar, Antibacterial and photocatalytic activities of sonochemically prepared ZnO and Ag-ZnO• , Journal of Alloys and Compounds, 2010, 508, 587-591.
13. C. Karunakaran, G. Abiramasundari, P. Gomathisankar, G. Manikandan, V. Anandi, Cu-doped TiO2 nanoparticles for photocatalytic disinfection of bacteria under visible light• , Journal of Colloid and Interface Science, 2010, 352, 68-74.
14. C. Karunakaran, R. Dhanalakshmi, P. Gomathisankar, Photomineralization of phenol on Al2O3. Synergistic photocatalysis by semiconductors• , Research on Chemical Intermediates, 2010, 36, 361-371.
15. C. Karunakaran, Sona Narayanan, P. Gomathisankar, Photocatalytic degradation of 1-naphthol by oxide ceramics with added bacterial disinfection• , Journal of Hazardous Materials, 2010, 181, 708-715.
16. C. Karunakaran, P. Gomathisankar, G. Manikandan, Preparation and characterization of antimicrobial Ce-doped ZnO nanoparticlesw for photocatalytic detoxification of cyanide• , Materials Chemistry and Physics, 2010, 123, 585-594.
17. C. Karunakaran, P. Anilkumar, P. Gomathisankar, Kinetics of Ag/TiO2-photocatalyzed iodide ion oxidation• , Monatshefte fur Chemie, 2010, 141, 529-537.
18. C. Karunakaran, P. Anilkumar, G. Manikandan, P. Gomathisankar, Solar-powered potentially-induced TiO2, ZnO and SnO2-catalyzed iodine generation• , Solar Energy Materials and Solar Cells, 2010, 94, 900-906.
19. C. Karunakaran, Semiconductor photoelectrocatalysis• in Eds. S. Kaneco, B. Viswanathan, H. Katsumata, Photo-Electrochemistry & Photo-Biology for the Sustainability, Bentham Science Publishers, 2010, 1, 129-166.
20. C. Karunakaran, R. Dhanalakshmi, P. Gomathisankar, G. Manikandan, Enhanced phenol-photodegradation by particulate semiconductor mixtures: interparticle electron-jump• , Journal of Hazardous Materials, 2010, 176, 799-806.
21. C. Karunakaran, M.P. Sujatha, P. Gomathisankar, Photoreduction of chromium(VI) on ZrO2 and ZnS surfaces, Monatshefte fur Chemie, 2009, 140, 1269-1274.
22. C. Karunakaran, R. Dhanalakshmi, P. Gomathisankar, Semiconductor-photocatalyzed degradation of carboxylic acids. Enhancement by particulate semiconductor mixture• , International Journal of Chemical Kinetics, 2009, 41, 716-726.
23. C. Karunakaran, P. Anilkumar, Solar-driven electrochemically assisted semiconductor-catalyzed iodide ion oxidation. Enhanced efficiency by oxide mixtures• , Central European Journal of Chemistry, 2009, 7, 519-523.
24. C. Karunakaran, R. Dhanalakshmi, P. Anilkumar, Photodegradation of carboxylic acids on Pr6O11 surface. Enhancement by semiconductors• , Chemical Engineering Journal, 2009, 151, 46-50.
25. C. Karunakaran, R. Dhanalakshmi, P. Anilkumar, Photooxidation of oxalic acid on Sm2O3. Synergism by semiconductors• , Catalysis Letters, 2009, 130, 222-226.
26. C. Karunakaran, R. Dhanalakshmi, P. Anilkumar, Photodegradation of phenol on Y2O3 surface. Synergism by semiconductors• , Journal of Hazardous Materials, 2009, 167, 664-668.
27. R. Kalaiselvan, A. Gedanken, P. Anilkumar, G. Manikandan, C. Karunakaran, Synthesis and characterization of rare earth orthovanadate (RVO4; R = La, Ce, Nd, Sm, Eu & Gd) nanorods/nanocrystals/nanospindles by a facile sonochemical method and their catalytic properties• , Journal of Cluster Science, 2009, 20, 291-305.

28. C. Karunakaran, R. Dhanalakshmi and P. Anilkumar, Degradation of carboxylic acids on Y2O3 surface under UV light. Synergism by semiconductors• , Radiation Physics and Chemistry, 2009, 78, 173-176.
29. C. Karunakaran and R. Dhanalakshmi, Selectivity in photocatalysis by particulate semiconductors• , Central European Journal of Chemistry, 2009, 7, 134-137.
30. C. Karunakaran and R. Dhanalakshmi, Substituent-effect on nano TiO2 and ZnO-catalyzed phenol-photodegradation rates• , International Journal of Chemical Kinetics, 2009, 41, 275-283.
31. C. Karunakaran and R. Dhanalakshmi, Phenol degradation on Pr6O11 surface under UV-A light. Synergistic photocatalysis by semiconductors• , Radiation Physics and Chemistry, 2009, 78, 8-12.
32. C. Karunakaran and R. Dhanalakshmi, Semiconductor-catalyzed degradation of phenols with sunlight• , Solar Energy Materials and Solar Cells, 2008, 92, 1315-1321.
33. C. Karunakaran and R. Dhanalakshmi, Photocatalytic performance of particulate semiconductors under natural sunshine-Oxidation of carboxylic acids• , Solar Energy Materials and Solar Cells, 2008, 92, 588-593.
34. C. Karunakaran and P. Anilkumar, Photooxidation of iodide ion on immobilized semiconductor powders• , Solar Energy Materials and Solar Cells, 2008, 92, 490-494

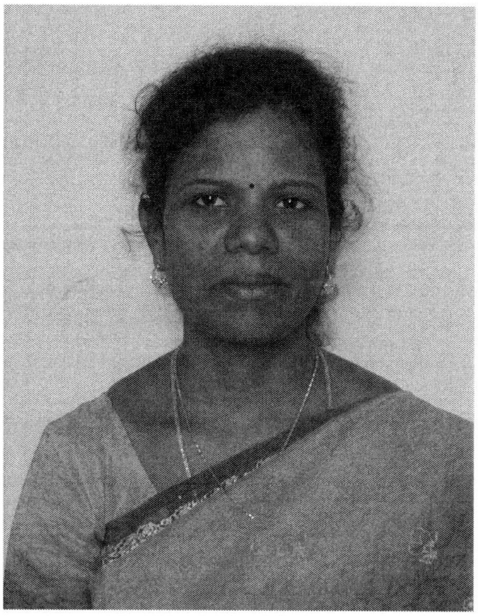

In: Chemistry Researcher Biographical Sketches
Editors: E. S. Huseynov and V. Babayev ISBN: 978-1-62100-771-5

Chapter 41

BIOGRAPHICAL SKETCH

Name:	Title:	Date of Birth:
Shahed U.M. Khan	Duquesne University	September 30, 1946

EDUCATION:

Institution and Location	Degree	Year Conferred	Scientific Field
Chemistry,	BS	1968	
Physical Chemistry	Ms	1970	
Electrochemistry and Photoelectrochemistry	Ph.D.	1977	

CONTACT POINTS:

Department of Chemistry and Biochemistry, Duquesne University, Pittsburgh PA 15282, USA

RESEARCH AND PROFESSIONAL EXPERIENCE:

Electrochemistry and Photoelectrochemistry

AWARDS AND HONORS:

Duquesne University president Award for Scholarship

Professional Appointments:

Associate Professor, Department of Chemistry and Biochem is try,

Publications during last three years

1. S. U. M Khan, M. Frites and W. B. Ingler Jr., ¡°Efficient Anode and Cathode Materials for Amorphous Silicon Solar Cell Driven all Solar Electrolysis of Water¡± Green, 1 (2011) 97-103.
2. M. Frites, W.B. Ingler Jr. and S.U. M. Khan, ¡°In Search of Efficient Anode and Cathode Materials for Double Junction Amorphous Silicon Solar Cell Driven all Solar Electrolysis of Water to Hydrogen and Oxygen gases¡± ECS Transactions, 28 (2010) 133-140
3. Y. A. Shaban and S. U. M. Khan, ¡°Efficient Photoelectrochemical Splitting of Water to H2 and O2 at Nanocrystalline Carbon Modified (CM)-n-TiO2 Thin Films ¡°, Solid State Phenomena, 162 (2010) 179-201.
4. M. Frites, W.B. Ingler and S. U. M. Khan, ¡°Photoelectrolysis of Water in Tj-a-Si Solar Cell Biased CM-n-TiO2| | Pt and in Monolithic Self-Driven n-TiO2-MnO2 coated Tj-a-Si | | Pt Photoelectrochemical Cells, ECS Transactions, 25 (2010) 73-82.
5. Y. A. Shaban and S. U. M. Khan, ¡°Efficient Photoelectrochemical Splitting of Water to H2 and O2 at Nanocrystalline Carbon Modified (CM)-n-TiO2 and CM-n-Fe2O3 Thin Films ¡° Int. J. Nanotechnology, 7 (2010) 68-98.
6. M. Frites, Y. A. Shaban and S. U. M. Khan, ¡°Iron Oxide (n-Fe2O3) Nanowire Films and Carbon Modified (CM)-n-Fe2O3 Thin Films for Hydrogen Production by Photosplitting of Water¡±, Intl. J. Hydrogen Energy, 35 (2010) 4944-4948.
7. M. Frites, W.B. Ingler Jr. and S. U. M. Khan, ¡°Carbon Modified (CM)-n-TiO2 for Efficient Water Splitting to Hydrogen and Oxygen Gases Under Solar Simulated Light Illumination, Material. Science & Technology Transaction, 9 (2009) 442-450.
8. M. Frites and S.U.M. Khan, ¡°Photoelectrochemical Splitting of Water to H2 and O2 at n-Fe2O3 Nanowires and Nanocrystalline Fe2O3 Thin Films¡±, ECS Trans. 19 (2009) 137-145.
9. M. Frites and S. U. M. Khan, ¡°Visible light Active Hydrogen Modified (HM)-n-TiO2 Thin Films for Photoelectrochemical Splitting of Water¡±, Electrochemistry Communication, 11 (2009) 2257- 2260.
10. Y. A. Shaban and S. U. M. Khan, ¡°Carbon Modified (CM)-n-TiO2 Thin Films for Efficient Water Splitting to H2 and O2 under Xenon Lamp Light and Natural Sunlight Illuminations¡±, J. Solid State Electrochem, 13 (2009) 1025-1036.

In: Chemistry Researcher Biographical Sketches
Editors: E. S. Huseynov and V. Babayev
ISBN: 978-1-62100-771-5

Chapter 42

BIOGRAPHICAL SKETCH

Name:	Title:	Date of Birth:
Renat Khaydarov	PhD	12/17/1974

EDUCATION:

Institution and Location	Degree	Year Conferred	Scientific Field
Tashkent State Univ, Physical department	M.Sc		

CONTACT POINTS:

Tel: +998 71 2893144, E-mail: renat2@gmail.com

RESEARCH AND PROFESSIONAL EXPERIENCE:

Dr. Khaydarov has 14 years experience in research. Much of his research has been in the study of synthesis methods and biomedical applications of silver and carbon nanoparticles. He has also participated in scientific projects on the development of new methods of water treatment.

AWARDS AND HONORS:

2002 - The second place in the competition of Young Scientists of Uzbekistan
2008 - The first place in the competition of Young Scientists of Uzbekistan
2007-2008 he was a Postdoctoral Fellow in University of Hannover, Germany working on his postdoctoral project entitled 'Antimicrobial effect of silver, copper and gold nano-particles in solutions and polymer materials'.

Professional Appointments:

Since 2001 he has been a senior research scientist for Institute of Nuclear Physics, Tashkent, Uzbekistan

<u>Publications during last three years</u>

1. Khaydarov, R.A., Khaydarov, R.R., Gapurova, O., Estrin, Y., and Scheper, T. (2009) "Electrochemical method of synthesis of silver nanoparticles", Journal of Nanoparticle Research, Vol. 11(5), pp. 1193-1200
2. Adlakha-Hutcheon G, Khaydarov R, Korenstein R et al, "Nanomaterials, nanotechnology: applications, consumer products, and benefits", in "Nanomaterials: Risk and Benefits", Series: NATO Science for Peace and Security Series C: Environmental Security, 2009, Springer, Netherlands, pp. 195-209
3. Khaydarov R.A, Khaydarov R.R., Cho S., (2009) "Natural disaster: prevention of drinking water scarcity" in: "Threats to Global Water Security" (Jones, J. et. al.: Eds.), pp. 381-384
4. Khaydarov R.A, Khaydarov R.R., Estrin Y., Cho S., Scheper T, and Endres C, "Silver nanoparticles: Environmental and human health impacts", in "Nanomaterials: Risk and Benefits" Series: NATO Science for Peace and Security Series C: Environmental Security, 2009, Springer, Netherlands, pp. 287-299.
5. Khaydarov R.R, Khaydarov R.A., Gapurova O., Estrin Y., Evgrafova S., Scheper T, Cho S.Y. "Antimicrobial effects of silver nanoparticles synthesized by an electrochemical method" in "Nanostructured Materials for Advanced Technological Applications", NATO Science for Peace and Security Series B: Physics and Biophysics, 2009, Springer, Netherlands, pp. 215-218
6. Khaydarov R.A, Khaydarov R.R., Gapurova O., "Remediation of Contaminated Groundwater Using Nano-Carbon Colloids", in Book "Nanomaterials: Risk and Benefits" Series: NATO Science for Peace and Security Series C: Environmental Security, 2009, Springer,Netherlands, pp. 219-225.
7. Rashid A. Khaydarov, Renat R. Khaydarov, Olga Gapurova <<Water purification from metal ions using carbon nanoparticle-conjugated polymer nanocomposites>>, Water Research Vol. 44 (2010), pp. 1927-1933
8. Rashid A. Khaydarov, Renat R. Khaydarov, Olga Gapurova, Yuri Estrin. <<A novel method of continuous fabrication of aqueous dispersions of silver nanoparticles>>, Int. J. Nanoparticles, No.3, 2010, pp. 77-91.
9. Rashid A. Khaydarov, Renat R. Khaydarov, Olga Gapurova <<Application of carbon nanoparticles for water treatment>>, <<Water Treatment Technologies for the Removal of High-Toxity Pollutants>>, Series: NATO Science for Peace and Security Series C: Environmental Security, Springer, Netherlands, (2010), pp. 253-258
10. R.A. Khaydarov, R.R. Khaydarov, O. Gapurova "Remediation of Metal Ions Contaminated Ground Water and Soil using Nanocarbon-Polymer Composition", <<Solid Waste Management and Environmental Remediation, Series: Environmental Remediation Technologies, Regulations and Safety>>, Editors: Timo Faerber and Johann Herzog (2010) Nova Science Publishers, USA.

11. Renat R. Khaydarov, Rashid A. Khaydarov, Yuri Estrin, Svetlana Evgrafova, Seung Y. Cho and Stefanie Wagner, "Antibacterial Applications of Silver Nanoparticles: Benefits and Human Health Risks", <<Silver Nanoparticles: Properties, Characterization and Applications, Series: Nanotechnology Science and Technology>>, Editors: Audrey E. Welles (2010) pp. 327-339, Nova Science Publishers, USA.
12. Rashid A. Khaydarov, Renat R. Khaydarov, Svetlana Evgrafova, Yuri Estrin <<Using silver nanoparticles as an antimicrobial agent", in Biodefence, Series: NATO Science for Peace and Security Series A: Chemistry and Biology, Springer, Netherlands, (2011), pp. 169-177.
13. Renat R. Khaydarov, Rashid A. Khaydarov, Olga Gapurova, Svetlana Evgrafova, Seung Y. Cho "The Preparation of Ag-Nanoparticle-Embedded Paints and their Antimicrobial Activity" in Paints: Types, Components and Applications, Series: Chemistry Research and Applications, NOVA Publishers, USA, (2011), pp. 156-172

In: Chemistry Researcher Biographical Sketches
Editors: E. S. Huseynov and V. Babayev

ISBN: 978-1-62100-771-5

Chapter 43

BIOGRAPHICAL SKETCH

Name:	Title:	Date of Birth:
Masashi Kijima	Institute of Materials Science, Graduate School of Pure and Applied Sciences, University of Tsukuba	09/12/1957

EDUCATION:

Institution and Location	Degree	Year Conferred	Scientific Field
Functional Polymers Tokyo Institute of Technology	PhD	1985	

CONTACT POINTS:

Asakusa 6-10-1, Taito, Tokyo111-0032, Japan

RESEARCH AND PROFESSIONAL EXPERIENCE:

Research on: Synthesis of Conjugated polymers and their application. Synthesis of nano-structured carbons.

Professional Appointments:

Research Associate (Tokyo Meteropolitan Univ), Lectuler & Assistant Professor (University of Tsukuba), Associate Professor (University of Tsukuba)

Publications during last three years

1. Bioresource Technology 102 (2011) 6279–6285.
2. *Maromolecules*, 42, 5946-5952 (2009)
3. *J. Mater. Chem.*, 18, 1037-1045 (2008)

In: Chemistry Researcher Biographical Sketches
Editors: E. S. Huseynov and V. Babayev

ISBN: 978-1-62100-771-5

Chapter 44

BIOGRAPHICAL SKETCH

Name:	Title:	Date of Birth:
Joong Kyun Kim	Pukyong National University, Korea	March 21, 1958

EDUCATION:

Institution and Location	Degree	Year Conferred	Scientific Field
PH.D. achieved from University of Missouri			

CONTACT POINTS:

599-1, Daeyeon- 3 Dong, Nam-Gu, Busan, Korea

RESEARCH AND PROFESSIONAL EXPERIENCE: 19 years

AWARDS AND HONORS: Professor with ten years of continuous service

Professional Appointments: Full professor

Publications during last three years

1) Treatment of fisheries wastes by biotechnology, Fisheries Sciences Research Organization, 2011.

2) Aerobic biodegradation of fish-meal wastewater from lab to large scale, ISBN: 1-60021-904-7, Nova Publishers, 2008.
3) Property of biodegraded fish-meal wastewater as a liquid-fertilizer, ISBN: 978-1-60456-483, Nova Publishers, 2008.

In: Chemistry Researcher Biographical Sketches
Editors: E. S. Huseynov and V. Babayev

ISBN: 978-1-62100-771-5

Chapter 45

BIOGRAPHICAL SKETCH

Name:	Title:	Date of Birth:
Jean Le Bras	Dr	23 October 1970

EDUCATION:

Institution and Location	Degree	Year Conferred	Scientific Field
ENSCP-Paris and Université Pierre et Marie Curie.	Engineering Diploma and MSc.degree (DEA)		
Organic chemistry	PhD	1998	

CONTACT POINTS:

E-mail: jean.lebras@univ-reims.fr

RESEARCH AND PROFESSIONAL EXPERIENCE:

CNRS fellow at Université de Reims Champagne Ardenne. Current interests: transition metal-catalysis with particular emphasis on oxidations, C-H activation, and valorization of agricultural by-products.

Publications during last three years

Pd-catalyzed Oxidative Coupling of 2-Alkylfurans with Olefins through C-H Activation: Synthesis of Difurylalkanes, E. Thiery, D. Harakat, J. Le Bras, J. Muzart *Organometallics* 2008, *27*, 3996-4004.

Palladium-catalyzed C-H bonds functionnalization in “Green Chemistry Research Trends”, J. Le Bras, E. Thiery, Editors: Jeffrey T. Pearlman, Hauppauge NY, *Nova Science Publishers, Inc.*, 2009, ISBN: 978-1-60692-092-3, pp 9-19.

Reactivity versus stability of oxiranes under Pd-catalyzed reductive conditions, E. Thiery, J. Le Bras, J. Muzart *Eur. J. Org. Chem.* 2009, 961-985.

Copper(I)-Catalyzed Synthesis of Substituted 2-Mercapto Benzimidazoles, S. Murru, B. K. Patel, J. Le Bras, J. Muzart *J. Org. Chem.* 2009, 74, 2217-2220.

Palladium-Catalyzed Dehydrogenative Coupling of Furans with Styrenes, C. Aouf, E. Thiery, J. Le Bras, J. Muzart *Org. Lett.* 2009, *11*, 4096-4099.

Palladium-Catalyzed Allylic Acyloxylation of Terminal Alkenes in the Presence of a Base, E. Thiery, C. Aouf, J. Belloy, D. Harakat, J. Le Bras, J. Muzart *J. Org. Chem.* 2010, *75*, 1771-1774.

Selective 5-exo-trig iodocyclization of N-tosyl-2-allylanilines in water, M. Batistatosa, M. A. Fousterisa, S. S. Nikolaropoulosa, J. Le Bras, J. Muzart *Lett. Org. Chem.* 2010, 7, 440-443.

From metal-catalyzed reactions with hydrosoluble ligands to reactions in and on water, J. Le Bras, J. Muzart *Curr. Org. Synth.* 2011, under press.

Low catalyst loading for the production of carboxylic acids from polysaccharides and hydrogen peroxide, C. Aouf, D. Harakat, J. Le Bras, J. Muzart, B. Estrine, S. Marinkovic, C. Ernenwein *ChemSusChem* 2010, *3*, 1200-1203.

Direct conversion of xylan into alkyl pentosides, F. Bouxin, S. Marinkovic, B. Estrine, J. Le Bras *Carbohydr. Res.* 2010, *345*, 2469–2473.

Intermolecular Dehydrogenative Heck Reactions, J. Le Bras, J. Muzart *Chem. Rev.* 2011, *accepté.*

In: Chemistry Researcher Biographical Sketches
Editors: E. S. Huseynov and V. Babayev
ISBN: 978-1-62100-771-5

Chapter 46

BIOGRAPHICAL SKETCH

Name:	Title:	Date of Birth:
Chu-Hsuan Lin	Professor	09 Jan 1981

EDUCATION:

Institution and Location	Degree	Year Conferred	Scientific Field
Electronics Engineering. of Nation Taiwan University	Ph.D		

CONTACT POINTS:

Institute of Opto-Electronic Engineering, National Dong Hwa University

AWARDS AND HONORS:

- Reviewer of Thin Solid Films
- Local Organizing Committee, 2010 Asia-Pacific Data Storage Conference (APDSC '10)
- Local Organizing Committee, International Symposium on Optical Memory 2010 (ISOM '10)
- Session Chair, 2008 International Electron Devices and Materials Symposium (IEDMS)
- Reviewer, 2008 International Electron Devices and Materials Symposium (IEDMS)

Professional Appointments:

- Assistant Professor in National Dong Hwa University

Publications during last three years

1. C.-H. Lin and C. W. Liu*, Metal-Insulator-Semiconductor Photodetectors,• sensors, Vol. 10, No.10, pp. 8797-8826, 2010.
2. C.-H. Lin*, Si/Ge/Si Double Heterojunction Solar Cells, Thin Solid Films, Vol. 518, No. 6, pp. S255-S258, 2010.
3. C.-H. Lin* and C. W. Liu, Metal-Oxide-Semiconductor SiGe/Si Quantum Dot Infrared Photodetectors with Delta Doping in Different Positions,• Thin Solid Films, Vol. 518, No. 6, pp. S237-S240, 2010.
4. W. S. Ho, Y.-H. Dai, Y. Deng, C.-H. Lin, Y.-Y. Chen, C.-H. Lee, and C. W. Liu*, Flexible Ge-on-polyimide detectors,• Appl. Phys. Lett., Vol. 94, No.26, p. 261107, 2009.
5. W. S. Ho, C.-H. Lin, T.-H. Cheng, W. W. Hsu, Y.-Y. Chen, P.-S. Kuo, and C. W. Liu*, Narrow-band metal-oxide-semiconductor photodetector,• Appl. Phys. Lett., Vol. 94, No.6, p. 061114, 2009.
6. C.-H. Lin, Y.-J. Yang, E. Encinas, W.-Y. Chen, J.-J. Tsai, and C. W. Liu*, Single Crystalline Film on Glass for Thin Film Solar Cells, J. Nanosci. Nanotechno. Vol. 9, No.6, p. 3622, 2009.
7. A. A. Abramov, C.-H. Lin, and C. W. Liu*, Fano Interference in the Quantum Well-Quantum Dot System, International Journal of Nanoscience, Vol. 7, No. 4/5, pp. 181-186, 2008.
8. C.-H. Lin, C.-Y. Yu, C.-C. Chang, C.-H. Lee, M. H. Liao, C.-Y. Peng, and C. W. Liu*, SiGe/Si quantum dot infrared photodetectors with delta doping,• IEEE Trans. Nanotech., Vol. 7, No. 5, pp. 558-564, 2008.

In: Chemistry Researcher Biographical Sketches
Editors: E. S. Huseynov and V. Babayev

ISBN: 978-1-62100-771-5

Chapter 47

BIOGRAPHICAL SKETCH

Name:	Title:	Date of Birth:
Cheng Liu	State Key Laboratory of Fine Chemicals, Department of Polymer Science and Materials, Dalian University of Technology, Dalian, People's Republic of China	15-11-1976

EDUCATION:

Institution and Location	Degree	Year Conferred	Scientific Field
Polymer Materials, Dalian University of Technology, Dalian, China Dissertation: Studies on synthesis, thermal properties and solubility of aromatic polyamides containing phthalazinone moiety	Ph.D./	Dec 2004	
Polymer Science, Dalian University of Technology, Dalian, China	B. S	July 1999	

CONTACT POINTS:

No.2 Linggong Road, Ganjingzi District, Dalian City, Liaoning Province, P. R. C., 116024

AWARDS AND HONORS:

2006 - Natural science award, Liaoning Province, China
2008 - Award for Technological Invention, China Petroleum and Chemical Industry Federation

RESEARCH AND PROFESSIONAL EXPERIENCE:

2005-2008 - Main research member of National High Technology Research and Development Program 863. Played key role in the development of high performance aromatic polyamides.
2007-2009 - Directed the National Natural Science Foundation of China.
2008- Present Responsible for the synthesis and properties of triaryl-s-triazine-based polymers.

Professional Appointments:

Vise director of Liaoning Province Engineering Center of High Performance Resins

Publications during last three years

1. Yu, G.; Liu, C.; Li, G.; Wang, J.; Jian, X., Thermal degradation kinetics of poly(aryl ether sulfone 1,3,5-triazine)s containing phthalazinone moieties. *Thermochimica Acta* 2011, 514, 51-57.
2. Yu, G. P.; Liu, C.; Wang, J. Y.; Xu, J.; Jian, X. G., Synthesis and characterization of poly(arylene ether s-triazine)s containing alkyl-, aryl- and chloro-substituted phthalazinone moieties in the main chain. *Polymer International* 2010, 59, (9), 1233-1239.
3. Yu, G. P.; Liu, C.; Wang, J. Y.; Li, G. H.; Han, Y. J.; Jian, X. G., Synthesis, characterization, and crosslinking of soluble cyano-containing poly(arylene ether)s bearing phthalazinone moiety. *Polymer* 2010, 51, (1), 100-109.
4. Yu, G.; Liu, C.; Wang, J.; Li, X.; Jian, X., Heat-resistant aromatic S-triazine-containing ring-chain polymers based on bis(ether nitrile)s: Synthesis and properties. *Polymer Degradation and Stability* 2010, 95, (12), 2445-2452.
5. Xiao, L. H.; Liu, C.; Liao, G. X.; Zhang, Y. A.; Jian, X. G., Synthesis and Characterization of Novel Copoly(aryl ether sulfone) Containing Phthalazinone and Biphenyl Moieties. *High Performance Polymers* 2010, 22, (3), 274-285.
6. Yu, G. P.; Wang, J. Y.; Liu, C.; Lin, E. C.; Jian, X. G., Soluble and curable poly(phthalazinone ether amide)s with terminal cyano groups and their crosslinking to heat resistant resin. *Polymer* 2009, 50, (7), 1700-1708.

7. Yu, G. P.; Liu, C.; Zhou, H. X.; Wang, J. Y.; Lin, E. C.; Jian, X. G., Synthesis and characterization of soluble copoly(arylene ether sulfone phenyl-s-triazine)s containing phthalazinone moieties in the main chain. *Polymer* 2009, 50, (19), 4520-4528.
8. Yu, G. P.; Liu, C.; Wang, J. Y.; Gu, T. S.; Jian, X. G., Synthesis, characterization and properties of heat-resistant and soluble poly(aryl ether)s containing s-triazine units in the main chain. *Polymer Degradation and Stability* 2009, 94, (7), 1053-1060.
9. Yu, G. P.; Liu, C.; Wang, J. Y.; Chu, C. B.; Jian, X. G., Synthesis of phenyl-s-triazine-based copoly(aryl ether)s derived from hydroquinone and resorcinol. *Polymer Degradation and Stability* 2009, 94, (11), 2065-2071.
10. Yan, C.; Zhang, S. H.; Liu, C.; Yang, D. L.; Yang, F. J.; Jian, X. G., Preparation, Morphologies, and Properties of Positively Charged Quaternized Poly(phthalazinone ether sulfone ketone) Nanofiltration Membranes. *Journal of Applied Polymer Science* 2009, 113, (3), 1389-1397.
11. Han, R. L.; Zhang, S. H.; Liu, C.; Wang, Y. T.; Jian, X. G., Effect of NaA zeolite particle addition on poly(phthalazinone ether sulfone ketone) composite ultrafiltration (UF) membrane performance. *Journal of Membrane Science* 2009, 345, (1-2), 5-12.
12. Wu, C.; Zhang, S.; Liu, C.; Yang, D.; Jian, X., Preparation, characterization and performance of thermal stable poly(phthalazinone ether amide) UF membranes. *Journal of Membrane Science* 2008, 311, 360-370.
13. Dong, L. M.; Liao, G. X.; Liu, C.; Yang, S. S.; Jian, X. G., SYNTHESIS AND CHARACTERIZATION OF POLY(PHTHALAZINONE ETHER NITRILE) COPOLYMERS WITH HYDROPHOBIC SURFACE. *Surface Review and Letters* 2008, 15, (5), 705-709.

In: Chemistry Researcher Biographical Sketches
Editors: E. S. Huseynov and V. Babayev
ISBN: 978-1-62100-771-5

Chapter 48

BIOGRAPHICAL SKETCH

Name:	Title:	Date of Birth:
Xiaoxuan Liu	Ph.D.	1957

EDUCATION:

Institution and Location	Degree	Year Conferred	Scientific Field
Chemistry engineering of polymer from the Dalian University of Science & Technology	B.Sc	1982.	
Polymer Chemistry and Physics, at the Institute of Polymer Science, Sun Yat-sen University, under the supervision of Prof. Yonglie Chen, where he worked on the reaction kinetics of reactive hindered amine light stabilizers (r-HALS) photopolymerization and applied to photostabilization *in situ* of UV-curable coatings	Doctorate	2001	

CONTACT POINTS:

Department of Polymeric Materials and Engineering
Faculty of Materials and Energy
Guangdong University of Technology

No.100, Waihuan Xi road
Guangzhou Higher Educational Mega Center
Panyu District, Guangzhou 51006, P. R. China.
Tel./Fax: +86-20-39322570(office), +86-20-39322165(Lab.)
Cell-Phone: +86-013600053965, +86-013424091088
E-mail address: p_xxliu@yahoo.com.cn or p-xxliu@gdut.edu.cn

RESEARCH AND PROFESSIONAL EXPERIENCE:

1982 to 1986- Assistant of chemistry in the Department of Optical Material Science, College of Optics and Fine Mechanics of Changchun, China. He was responsible for the research on the optical plastics and plastic optical fibers, and studied on the gradient-refractivity of optical materials.

1986 to 2006 - held appointments at the Shantou University which received foundation donated by Mr. Li Ka-shing, a generous support from the distinguished philanthropist and international entrepreneur.

Aug 1995 to Jul 1996 - joined in the Department of Chemistry, Tsinghua University. He held a visiting scholarship supervised by Prof. Xinqi Song, he was collaborative with Ph.D. Yong Qiu to study on the organic luminescence LED and got the important achievement as the basic research for the future.

He was a lecture (1986~1994), associated professor (1995~2003), and professor (2004~2006) as a faculty of Shantou University.

2004 University professor, the highest distinction faculty can achieve at Shantou University.

2006 to present, - served as head of the Department of Polymeric Materials and Engineering in the Faculty of Materials and Energy, Guangdong University of Technology, and supervisor for Ph.D. & M. Sc. students major in polymer photochemistry and UV-curable materials.

Since 2001 - researched focus on the controlled radical photopolymerization, including NMP and ATRP, three projects supported by National Natural Science Foundation of China. At the same time, he formed a research consortium with various industrial corporations to expand the technique and application of photopolymerization, including UV-curable, photo-grafting and accelerate the transfer of this technology to different commercial applications. Xiaoxuan Liu has 4 active CN patent technologies and applications. One of the respective educators in the field of polymer photochemistry, Xiaoxuan Liu has 1 current doctoral student and 2 postdoctoral fellows. He has mentored more than 34 undergraduates, graduate students. Since 1986, he has authored 1 book chapter and more than 70 peer-reviewed scientific papers.

AWARDS AND HONORS:

Numerous awards for his work, including the scholarship for outstanding students in 1981, at the Dalian University of Science & Technology. He was also honored the Excellent Doctoral Dissertation Award in 2001, at the Sun Yat-sen University; From 1990 to 1994, He received First-Prize Winner as the Working Foundation Grant by Mr. Kashing Li (Hong Kong) for five years continuously. He was also honored the Awards for an Excellent Thesis, First Academic Annual Meeting of Youth Proceedings in 1992, China Association for Science and Technology. During Shantou University, he was received many awards such as Second-Prize Winner in Awards for an Excellent Teaching in 1992, Awards for Outstanding Director in Lab. Management in 1995; First-Prize Winner in Awards for an Excellent Research of Science in 1997, and Awards for Achievement in Science Research and Excellent Teaching Achievement, in 2002 & 2003 respectively. In 2006 The Third-class Award for Science & Technology Achievement of Shantou Government, China. In 2010, he was elected as a fellow of the China Society of Chemistry and as international fellow of the Royal Society of Chemistry (RSC) UK and RadTech North USA. From 2010 to 2015, he was engaged for a technological adviser of Academic Communication for Dongguan Entry-exit Inspection & Quarantine of the People's Republic of China (CCIQ). At same times, he was elected as an academic membership of RadTech China Association, 2010~present.

Professional Appointments:

Professor in Dep. Polym. Mater. & Eng. at the Faculty of Materials and Energy, Guangdong University of Technology, China, is an respective polymer chemist who is highly regarded for his vision, his supervising for many Ph.D. & M. Sc. graduates in education and his many collaborative and research efforts that have yielded significant innovations in polymer photochemistry. His research has focused on the nitroxide mediated radical polymerization (NMP) with emphasis on the design, synthesis, and characterization of novel nitroxide derivatives and initiator with unusual structures. He is focusing on the development of new NMP photo-polymerization techniques and novel approaches to the functional materials that have renovated the new way to prepare polymer with narrow polydispersity and block copolymers are made.

Publications during last three years

Journal Papers

1. Yao Yongping, Liu Xiaoxuan. *Research Development of UV-curing Waterborne Coating*, Journal of Coating Industry (in Chinese), 2011, 3 (to be accepted)
2. Zhang Yongtao, Liu Xiaoxuan. *Research Development of Vacuum Plating Coating*, Journal of Coating Industry (in Chinese), 2010, 12 (to be accepted)
3. Liu Xiaoxuan, Zhang Yongtao, Cui Yanyan, Yang Weisheng, Guo Peifei, Dong Zhixian. *Synthesis of Porous Hydrogels of NIPAM Copolymer by Reversible Addition Fragmentation Chain Transfer Photopolymerization*, Acta Polymerica Sinica (APS) (in Chinese), 2011, 5: 802-808 (SCI accepted)
4. Yang Xuemei, Cui Yanyan, Hong Peng, Dong Zhixian, Liu Xiaoxuan. *Research on Sunthesis of Glycidyl Methacrylate*, J. Guangdong University of Technology (Natural Science Edition) (in Chinese), 2010, 27(4): (in press)

5. Guodong Ye, Wenbo Kan, Jianwen Yang, Zhaohua Zeng, Xiaoxuan Liu, *A study on hybrid oligomeric type II photoinitiator*, *e*-Polymers, 2010, no. 138 （SCI accepted）
6. Zhang Ting, Chen Jingdong, Liu Xiaoxuan, *Study on the Reactivity Ratios and Sequence-length Distribution of Photo-induced Copolymerization of APMP and MMA in Solution*, J. IANGXI Science (in Chinese), 2009, 27(1): 116-119
7. Yang Xuemei, Chen Fulin, Liu Xiaoxuan, Zhou Yanhao. *Recent Development of the Peroxide Vulcanization of the Organic Tin Applied to Rubber*, Journal of Special Purpose Rubber Products (in Chinese), 2009, 30(3): 79-84
8. Zhan Shaotong, Liu Xiaoxuan *et al.*, Stat Reverse Expansion Polymerization to synthesis Sodium Polyacrylate, Journal of Guangdong Chemical Industry (Natural Science Edition) (in Chinese), 2009, 37(1): 12-14
9. Liu Xiaoxuan, Wang Hongbo, Chen Bing, Jing Yanni, Wu Guangguo. Photochemical Synthesis and Characterization of Poly(N,N'-dimethylacrylamide-*co*-n-isopropylacrylamide) Hydrogels with Fast Responsibility, Acta Polymerica Sinica (APS) (in Chinese), 2007, 9: 850-856 (SCI accepted, IDS Number: 215 ET)
10. Liu Xiaoxuan, Bai Yingkun, Jing Yanni , Wang Hongbo, Chen Bing, Wu Guangguo. *The Anchoring of the Photoinitiator and the Photografting Polymerization of MMA on the surface of Nanometer Silica Particles*, Acta Polymerica Sinica (APS) (in Chinese), 2007, 1038-1044 (SCI accepted)
11. Xiaoxuan Liu, Yanni Jing, Yingkun Bai, Guangguo Wu, Controlled/"living" *photopolymerization of methyl methacrylate in Miniemulsion Mediated by HTEMPO*, Chem. J. Chinese Universities[J] (in Chinese), 2007, 28(4): 774-778 （SCI收录，IDS Number: 159NZ）
12. Liu Xiaoxuan, Wang Hongbo , Bai Yingkun, Wu Guangguo. *Photochemical Synthesis and Characterization of P(DMAA-co-MMA)/Nano-SiO2 Complex Hydrogels*, Journal of Photographic Science and Photochemistry(in Chinese), 2007, 25(6): 436-444 (EI accepted)
13. Lin Shaoyu, Gao Wenhua, Liu Xiaoxuan, Chen Yaowen. *Determination of Trace Chromium(VI) by Catalytic Spectrofluorimetry Using Pyronine Y-H2O2-Cr(VI) System*, Spectroscopy and Spectral Analysis, 2007, 25(1): 113-115
14. Liu Xiaoxuan, Chen Bing, Bai Yingkun, Wang Hongbo, Jing Yanni, Wu Guangguo. *Photo-stabilization in situ of Reactive-hindered Amine Derivatives (r-HALS) Applied in UV-curing Coatings*, Journal of Liaodong University (Natural Sciences) (in Chinese), 2007, 14(1): 1-6.
15. Liu Xiaoxuan, Deng Xianghua, Ren Yae, Wu Guangguo, *Study on the reactivity ratios and sequence-length distribution of photo-induced copolymerization of MA and AATP in solution,* Acta Polymerica Sinica (APS) (in Chinese), 2006,1: 118-123 (SCI accepted)
16. Liu Xiaoxuan, Ren Yae, Zhang Ting, Han Mei, Wu Guangguo. *Controlled/Living Photopolymerization of Methyl Methacrylate Mediated by Bimolecular Systems for HTEMPO/2,2,6,6-Tetramethyl-Piperidinone*, Chem. J. Chinese Universities[J] (in Chinese), 2006, 27(1): 192-195 (SCI accepted)

17. Liu Xiao-Xuan, Han Mei, Duan Lun-Yong, Wu Guang-Guo, *Controlled Radical Photopolymerization of n-Butyl Methacrylate in Bluk Mediated by HTEMPO*, Chinese Chemical Letter, 2006, 17(8): 1109-1112 (SCI accepted)
18. Guangguo Wu, Yuanpei Li, Mei Han, Xiaoxuan Liu *Novel Thermo-sensitive Membranes Prepared by Rapid Bulk Photo-grafting Polymerization of N,N-diethylacrylamide onto the Microfiltration Membranes of Nylon*, Journal of Membrane Science, 2006, 283: 13-20 (SCI accepted)
19. Guangguo Wu, Yuanpei Li, Enyi Ou, Mei Han, Xiaoxuan Liu* , *Preparation and Characterization of N,N-diethylacrylamide and Acrylic Acide Binary Monomers-grafted Polypropylene Films with Temperature and pH-Sensitivities*, Acta Polymerica Sinica (APS) (in Chinese), 2006, 9: 1022-1028 (SCI accepted)
20. Xiaoxuan Liu, Yae Ren, Guangguo Wu Xiahong Zhang, Jiangwen Yang,Yonglie Chen, *Determining of monomer reactivity ratios and sequence structure for photo-copolymerization of methyl methacrylate with 4-methacryloyl-1,2,2,6,6-pentamethyl-piperidinyl*, the 40th International Symposium on Macromolecules (The Paris World Polymer Congress, MACRO 2004), sponsored by IUPAC, e-Polymers, 2005, on. E_002
21. Guangguo Wu, Xianghua Deng, Xiaoxuan Liu, *Study on photostabilization in-situ of UV-curing coatings of aromatic polyurethane acrylate*, the 40th International Symposium on Macromolecules (The Paris World Polymer Congress, MACRO 2004), sponsored by IUPAC, e-Polymers, 2005, on. E_002
22. Xiaoxuan Liu, Xiahong Zhang, Xinghong Zhang, Guangguo Wu, Jianwen Yang, Laixing Pang, Zhaohua Zeng And Yonglie Chen, *Controlled / 'Living' Free Radical Photopolymerization of 4-Methacryloyl -1,2,2,6,6- pentamethylpiperidenyl,* Journal of Polymer Science, Part A: Polym. Chem., 2004, 42: 2659-2665 (SCI accepted, IDS number: 822GO)
23. Bei Yiling, Liu Xiaoxuan, Zhang Hongxing, Wu Guangguo, *Study on the Microstructures and Thermal Stability of MA/ATMP Copolymers*, Material Engineering(in Chinese), 2004, 7: 14-17 (EI accepted)
24. Liu Xiaoxuan, Deng Xianghua, Guo Xu, Ren Ae, Wu Guangguo, *Reaction Kinetics of Photoinduced Bulk Polymerization of 4-Acrylamido-2,2,6,6-Tetramethyl-Piperidine in Melting State*, Journal of Photographic Science and Photochemistry(in Chinese), 2004, 22(4): 277-286(EI accepted)
25. Liu Xiaoxuan , Ren Yae, Guo Xu, Zhang Ting, Han Mei, Guangguo Wu, *Study on UV-curing Kinetics of Epoxy Acrylate/Allyl Ether System*, Journal of South-Central University for Nationalities, Nature Science Edition, 2004, 23(4): 1-4
26. Xiaoxuan Liu, Xinghong Zhang, Guangguo Wu, Jianwen Yang, Zhaohua Zeng And Yonglie Chen, *A Kinetic Evidence for the Nitroxyl Radicals Recycling Mechanism in the Photo-stabilizing Process of HALS*, Chinese Chemical Letters, 2003, 14(10): 1085-1088 (SCI accepted, IDS number: 733DA)
27. Liu Xiaoxuan , Linghai Xie, Weiqun Zhang, Yanping Xu, Guangguo Wu, Laixing Pang, Jianwen Yang, Zhaohua Zeng And Yonglie Chen, *Surface modification of LLDPE film with MPMP by photo-grafting copolymerization,* Chemical Journal on Internet, 2003, 5(7):58-66

28. Linghai Xie Xiaoxuan Liu Weiqun Zhang Yanping Xu Guangguo Wu, *Studies on the Surface Modification of LLDPE Film by the Photo-grafting of MTMP*, Journal of Shantou University, Nature Science Edition, (in Chinese), 2003,18 (4): 31-38
29. Xiaoxuan Liu Xu Guo Ting Zhang Xianghua Deng Guangguo Wu, *The kinetic study on epoxy acrylate/vinyl photocuring system initiated by sulfonium salt*, Journal of Shantou University, Nature Science Edition, (in Chinese), 2003,19(1): 40-45
30. LIU Xiao-xuan, ZHANG Xing-hong, YIN Qiang and WU Guang-guo, *Reactive-HALS III: Photopolymerization Kinetics of ATMP*, Journal of Functional Polymers (in Chinese), 2003,16(1): 13-20
31. Zhang Xinghong, Liu Xiaoxuan, Yin Qiang and Wu Guangguo, *Study on Photo-polymerizing Kinetics of MTMP in Melting State*, Journal of Shantou University, Nature Science Edition, (in Chinese), 2003, 18(1): 31-38
32. Xiaoxuan LIU, Jianwen YANG and Yonglie CHEN, *Reactive-HALS I: Synthesis, Characterization, Copolymerization Reactivity and Photo-polymerizing Performance Applied in UV-curable Coatings*, Polymers for Advanced Technologies (PAT), 2002, 13: 247-253 (SCI and ISTP accepted, SCI IDS number: 527RH)
33. Xiaoxuan LIU, Guangguo WU, Yu CHEN, Jianwen YANG, Zhaohua ZENG and Yonglie CHEN, *Reactive-HALS II: Photo-polymerizing and UV-curing Kinetics of APMP and MPMP*, Chemical Research in Chinese Universities, 2002, 18(3): 159-160
34. LIU Xiaoxuan, YANG Jianwen, HUANG Jintao and CHEN Yonglie, *Study on the UV-curing Kinetics and Photostabilization of Polyurethane Acrylate Coatings,* Journal of Functional Polymers (in Chinese), 2001,14(4): 387-392
35. LIU Xiaoxuan, GAO Wenhua and ZHUANG Xiaolin, *Recycling and Utilizing of Processing Scraps of PVDC Coatings*, «Study on the Strategies for the Population, Resources, Environment and Continuous Development of China» (Chinese), published by China Environment Science Press, Beijing, 2000, p. 4085 (Awards for an Excellent Paper of the Edition Committee, Certification No. 03175)
36. LIU Xiaoxuan and CHEN Yonglie, *Recent Developments in Photostabilization of Polymer Coatings,* Polymer Bulletin (in Chinese), 1999, 4: 57-66
37. Yong Qiu, Deqiang Zhang, Zhiliang Xie and Xiaoxuan Liu, *Green Light-Emitting Organic Electroluminescent Devices with a New Fluorescent Dye Dispersed in Poly (N-vinylcarbazole) Emitter Layer*, Chinese Chemical Letter, 1997, 8(1): 79-82 (SCI accepted, IDS number:)
38. Qiu Yong, Zhang Deqiang, Liu Xiaoxuan and Xie Zhiliang, *Construction and Evaluation of Organic Electro-luminescent Devices,* Journal of Tsinghua University, Nature Science Edition, (in Chinese), 1997, 37(12): 69(EI accepted)
39. Xiaoxuan LIU and Kunjiu HU, *Studies on Photo-protection and Photo-oxidation in Polystyrene (PS) with Various Wavelength Region*, Polymers For Advanced Technologies, 1996, 7(2): 117-121 (SCI, EI and ISTP accepted, SCI IDS number:)
40. Xiaoxuan Liu, Wenhong Lin and Hao Lin, *Study on the Photostabilization of Polystyrene (PS) Film by PDS*, «China Youths Processings Choice for Science and Technology» (Chinese), China Sci. and Tech. Press, Beijing 1994, p.482-486
41. Xiaoxuan Liu, Hao Lin, Wenhong Lin, Jinbiao Xie and Zhibing Lu, *Study on the Early-Stage Characterization and Spectral Behavior of Polysty-rene (PS) Film during Photooxidation Process,* (Awards for Excellent Thesis, China Association for Science and Technology First Academic Annual Meeting of Youths Proceedings) «China

Youths Proceedings Choice for Science and Technology (Chinese) », China Sci. and Tech. Press. Beijing 1994, p.487-491

42. Xiaoxuan Liu and Zhigang Lu, *Study on UV Radiation Measurement with a Domestic Polysulphone*, Journal of Shantou University, Nature Science Edition (in Chinese), 1990, 5(2): 116-121
43. Yengtai Jin, Yufan Sun, Xiaoxuan Liu, Xingmin Li and Jun Qayang, *Investigation on Polymerization of Conjugation of Conjugated Dienes with the Homogeneous Rare Earth Catalyst,* Journal of Chemical Research and Molecule Science (Chinese), 1984, (2): 247-252, (CA, 101(16): 131149G)

Presentations

1. Liu Xiaoxuan, Hong Peng, et al., Synthesis and Characterization of the (n-BA)-b-St Copolymers by Photopolymerization of NMP/ATRP Technique using Macroinitiator in the Microemulsion. MACRO2010, 43rd IUPAC World Polymer Congress, 11-16 July 2010, SECC, Glasgow, UK
2. Liu Xiaoxuan, GUO Peifei, et al., Controlled Photopolymerization of n-Butyl Methacrylate in Microemulsion Mediated by HO-TEMPO, MACRO2010, 43rd IUPAC World Polymer Congress, 11-16 July 2010, SECC, Glasgow, UK
3. Liu Xiaoxuan, Hong Peng, Duan Lunyong and Li Yuanpei, Preparation of Water-borne Polyurethane-acrylate (PUA) and Application to UV-curing Coatings on the Package of Paper, RadTech UV & EB Technology Expo & Conference 2010, 23-26 May 2010, Baltimore, MD, USA.
4. Xiaoxuan Liu, Yae Ren, Guangguo Wu Xiahong Zhang, Jiangwen Yang,Yonglie Chen, Determining of monomer reactivity ratios and sequence structure for photo-copolymerization of methyl methacrylate with 4-methacryloyl-1,2,2,6,6-pentamethyl-piperidinyl, the 40th International Symposium on Macromolecules (The Paris World Polymer Congress, MACRO 2004), July 4-9, 2004, Paris, France.
5. Guangguo Wu, Xianghua Deng, Xiaoxuan Liu, Study on photostabilization in-situ of UV-curing coatings of aromatic polyurethane acrylate, the 40th International Symposium on Macromolecules (The Paris World Polymer Congress, MACRO 2004), July 4-9, 2004, Paris, France.
6. Xiaoxuan Liu, Guangguo Wu, Yu Chen, Jianwen Yang, Zhaohua Zeng And Yonglie Chen, Reactive-HALS II: Photo-polymerizing and UV-curing Kinetics of APMP and MPMP, Proceedings of the 12th International Symposium on Fine Chemistry and Functional Polymers, Agu.5-8, 2002, Lanzhou, China, Chemical Research in Chinese Universities, 2002, 18(3): 159-160
7. Xiaoxuan Liu, Jianwen Yang And Yonglie Chen, Reactive-Hals: Synthesis, Characterization and Photopolymerization Kinetics, Proceedings of International Conference on New Trends for Functional Polymer, May 8-13, 2000, Huangshan, China, p.76-80
8. Xiaoxuan Liu, Jintao Huang, Jianwen Yang And Yonglie Chen, The Photostabilization of r-HALS Applied in UV-curable Coatings, Proceedings of 10th International Symposium on Fine Chemistry and Functional Polymers, Sept. 20-23, 2000, Jinan, China, p.122-123

9. Xiaoxuan Liu, Jianwen Yang And Yonglie Chen, Synthesis, Characterization and Kinetic Performance of Bifunctional Stabilizer HALS in UV-curable Coatings, Proceedings of 6th Pacific Polymer Conference (PPC-6), Dec 7-11, 1999, Guangzhou, China, p.234
10. Xiaoxuan Liu and Kunjiu Hu, Studies on Photoprotection and Photooxidation in Polystyrene (PS) with Various Wavelength Region, Proceedings of the 5th International Symposium on Fine Chemistry and Functional Polymers, Oct.12, 1994, Jinan, China (ISTP accepted), p. a326

Research Projects Sponsored (in recent years)

1. Jan. 2003 ~ Dec. 2005, Theory and Application of photopolymerization, the Group Funding of Polymer Science by the "National 211 progress" of Shantou University, Jan. 2005 ~ Dec. 2007, *Studies on the Controlled/living Photo-polymerization and Kinetic Mechanism of Reactive Hindered Amine Piperidine Derivatives*, The project funding by National Natural Science Foundation of China (NNSFC, No. 20474036)
2. Jan. 2006 ~ Dec. 2008, *Studies on Reaction Kinetics of Controlled/living Photo-polymerization in Microemulsion*, The project of Doctoral funding by Guangdong University of Technology July 2008 ~ July 2009, *Study on Preparation and Application of novel UV-curable resin for Vacuum Plating Coatings*, The project funding by Guangdong SHENZHAN Industry Co. Ltd.
3. Aug. 2008 ~ Ayg. 2009, *Study on Preparation and Application of novel UV-curable Transfer Printing Film and Adhesives*, The project funding by Shantou DONGTIAN Transfer Printing Film Co. Ltd.
4. Oct. 2008 ~ Dec. 2009, *Study on Preparation and Application of novel UV-curable resin for Plastics*, The project funding by Dongguan BEITELI New Materials Co. Ltd.
5. March. 2011 ~ Sep. 2011, Preparation and trying production of *Novel* UV-curable Waterborne Resins, the Projects funding supported by Guangzhou KEYUAN New Material Co. Ltd.
6. Jan. 2009~ Dec. 2011, *Studies on the Controlled/living Photo-polymerization and Kinetic Mechanism in Microemulsion and Synthesis of "core-shell" Polymer Nano-particle by Photochemistry*, The project funding by National Natural Science Foundation of China (NNSFC, No. 20874022)
7. Jan. 2010 ~ Dec. 2012, *Studies on the UV-curing Kinetics Radial/Cation Hybrid System of the epoxy acrylate and Preparation of UV-curable Heat Press Adhesives*, The project of Doctoral funding by National Ministry of Education (NME, No. 20094420110006)
8. Jan. 2010 ~ Dec. 2012, Preparation and trying production of UV-curable Waterborne Vacuum Plating Coatings, the Projects funding supported by Foundation of Guangdong Province Government of China. (FGPGC, No. 20094420110006)

9. Jan. 2009 ~ Dec. 2011, Study on the Photopolymerization Theory and Applied to Prepare Advanced Engineering Materials, the Group Funding of Polymer Science by the "National 211 progress" of Guangdong University of Technology,

PATENTS

1. CN Patent No. ZL 200410077230.4, 2008 Solution recycling method of processing scraps of PVDC package film
2. CN Patent No. ZL 200410077398.5, 2007 Formulation of waterborne reactive-HALS and applied to photo-stabilization *in situ* of the UV-curable coatings
3. CN Patent No. ZL 200410077372.0, 2007 Formulation of reactive-HALS and applied to photo-stabilization *in situ* of the UV-curable coatings
4. CN Patent No. ZL 200610035457.1, 2006 Preparation method of novel UV-curable waterborne polyurethane acrylate resin

In: Chemistry Researcher Biographical Sketches
Editors: E. S. Huseynov and V. Babayev
ISBN: 978-1-62100-771-5

Chapter 49

BIOGRAPHICAL SKETCH

Name:	Title:	Date of Birth:
Yulia Sergeevna Lukina	D. Mendeleyev University of Chemical Technology of Russia, Moscow, Russia and N. Priorov Institute of Traumatology and Orthopedy of Russia, Moscow, Russia	15.11.1978

EDUCATION:

Institution and Location	Degree	Year Conferred	Scientific Field
D. Mendeleyev University of Chemical Technology of Russia, Moscow, Russia	Ph.D. Cand. of Tech. Sciences	2011	

CONTACT POINTS:

Sovetskaya str., 3, apartm.123, Noginsk, Moskow oblast 142400, Russia

RESEARCH AND PROFESSIONAL EXPERIENCE:

Researches in the field of biomaterials of 8 yearsï

Professional Appointments:

Researcher

Publictions during last three years

1. Lukina Yu. S., *Sventskaya* N. V. Biocomposicionnii material na osnove dicalciiphosphata digidrata // Steklo i keramica. – 2010. – â„– 11. –. " S. 23 – 26 *(in Russian).* Yu. S. Lukina and N. V. Sventskaya. Biocomposite material based on dicalcium phosphate dihydrate // Glass and Ceramics. – 2011. - V. 67, â„– 11 – 12. 2. P. 354 – 357. DOI: 10.1007/s10717-011-9297-9.
2. Sivkov S. P., Lukina Yu. S. Sintes i issledovanie svoistv inekcionnih kalciiphosphatnih brushitovih cementov // Tehnika i tehnologiya silikatov. - 2010. – Ð¢. 17, â„–4.. – S. 10 – 16 *(in Russian).*
3. Lukinaina Yu. S. Svoistva biocementa na osnove dicalciphosphata digidrata, modificirovannogo dobavkami, povishauschimi plastichnost // Uspehi v himii i himicheskoi tehnologii: Sb. nauch. tr. XXIII Mejd. konf. molodih uchenih po himii i him. tehn. - Ðœ. – 2009. – Ð¢. XXIII, â„– 7. – S. 26S. 26 – 30 *(in Russian).*
4. Artuhov Ð• . Ð• ., Shtilman M. I., Pashkova L. I., Lukina Yu. S. Composicionnie biodegradiruemie materiali dlya regeneracii kostnoi tkani // Biomateriali v medicine: Tez. dok. Vseross. sovesch. – M., 2009. – S. 11 – 12 *(in Russian).*
5. Zaitsev V. V., Lukina Yu. S., Lunin V. G., Karyagina A. S., Semihin A. S. Kostnie morfogeneticheskie belki (Ð'ÐœÐ) na nositele iz kalciiphosphatnogo cementa dlya regeneracii kostnoi tkani // // Biomateriali v medicine: Tez. dok. Vseross. sovesch. - Ðœ., 2009. – Ð¡. 35 – 36 *(in Russianian)*
6. Semihin A. S., Lunin V. G., Karyagina A. S., Zaitsev V. V., Lukina Yu. S. Ispolzovanie Ð'ÐœÐ 2 dlya sozdaniya composicionnih preparatov novogo pokoleniya // Biomateriali v medicine: Tez. dok. Vseross. sovesch.– M., 2009. – S. 61 – 62 *(in Russian).*
7. Lukina Yu. S., Berchenko G. N., Sivkov S. P., Gavrushenko N. S., Semihin A. S., Lavrova N. V., Poletaeva N. N., Zaitsev V. V., Karyagina A. S., Lunin V. G. Predvaritelnie dannie po razrabotke rezorbiruemogo kalcii-phosphatnogo cementa na osnove brushita, soderjaschego recombinantnii kostnii morfogeneticheskii belok (rhÐ'ÐœÐ -2) // Primenenie iskusstvennih kalcii-phosphatnih biomaterialov v travmatologii i ortopedii: Sb. rab. Vseross. nauchno-praktich. konf. –M. – 2010. – S. 7 *(in Russian).*
8. Zaitsev V. V., Karyagina A. S., Lunin V. G., Semihin A. S., Gromov A. V., Lavrova N. V., Poletaeva N. N., Galushkina Z. M., Lukina Yu. S., Zolotuhina M. S., Soboleva L. A., Berchenko G. N., Fronchek A. V., Eskin N. Ð• . Eksperimentalnaya ocenka osteoinduktivnosti implantatov na osnove biologicheskogo i sinteticheskogo matriksov v kachestve nositelya i recombinantnogo kostnogo morfogeneticheskogo belka (rhÐ'ÐœÐ -2) // IX sezd travmatologov – ortopedov: Tez. dokl.: v 3 t. – Saratov, 2010. – " T.1. – S. 1105 – 1106 (in Russian).
9. Lukina Yu. S., Sivkov S. P., Zaitsev V. V., Gromov A. V., Semihin A. S., Karyagina A. S., Lunin V. G. Issledovanie vozmojnosti rentgenologicheskogo kontrolya biodegradacii kalciiphosphatnogo cementa // Novie tehnologii sozdaniya i primeneniya biokeramiki v vosstanovitelnoi medicine: Mater. Vserr. nauchno-practich. konf. – Tomsk, 22010. – S. 93 – 99 *(in Russian).*

In: Chemistry Researcher Biographical Sketches
Editors: E. S. Huseynov and V. Babayev

ISBN: 978-1-62100-771-5

Chapter 50

BIOGRAPHICAL SKETCH

Name:	Title:	Date of Birth:
Maria Virginia Luna	Laboratorio de Fisiología Vegetal, Departamento de Ciencias Naturales, Fac. de Cs. Exactas, Universidad Nacional de Río Cuarto, Ruta 36 Km 601, 5800 Río Cuarto, Argentina.	August 29th, 1953

EDUCATION:

Institution and Location	Degree	Year Conferred	Scientific Field
Biological Sciences	PhD		

CONTACT POINTS:

1441 Jujuy Street, Rio Cuarto City, Córdoba State, Argentina

RESEARCH AND PROFESSIONAL EXPERIENCE:

30 years of experience as professor and scientist in Plant Physiology, with special emphasis in Stress Physiology, Plant Hormones, Plant-Microbe interactions.

AWARDS AND HONORS:

Senior Resercher of the National Council of Science and Technology (CONICET), Ministry of Sience, Argentina.

Member of the Advisors Comission in Agronomy for CONICET.

Member of the Academic Comitee for the Binational Doctorate in Science and Technology between Brasil and Argentina.

Member of the Editorail Board of Eurasian Journal of Soil Sciences.

Member of the Directory Board of the National Society of Plant Physiology.

Professional Appointments:

Associate Professor in Plant Physiology, Department. of Natural Sciences,
National University of Rio Cuarto, Argentina.

Publications during last three years

1. Luna V., Llanes A., Sosa L., Reginato M., Reinoso H. 2008. Differential effects of sodium salts on the germination of a native halophytic species from South America: Prosopis strombulifera (Lam.) Benth. In: Abdelly et al. (Eds): Biosaline Agriculture and High Salinity Tolerance: pp. 81-90. BIRKHAUSER Verlag. Switzerland. ISBN: 978-3-7643-8553-8.
2. Cassán F., Maiale S., Masciarelli O., Vidal O., Luna V., Ruiz O. 2008. Cadaverine production by Azospirillum brasilense and their possible role in plant growth promotion and osmotic stress mitigation. European J. of Soil Biology, 8 (3): 1-8.
3. Cassán F., Perrig D., Sgroy V., Masciarelli O.,Penna C., Luna V. 2008. Azospirillum brasilense Az39 and Bradyrhizobium japonicum E 109 promote seed germination and early seedling growth, independently or co-inoculated in maize (Zea mays L.) and soybean (Glycine max L.). European J. of Soil Biology, 8 (3): 9-17.
4. Andrea Andrade, Oscar Masciarelli, Sergio Alemano, Virginia Luna, Guillermina Abdala. 2008. Tomato mutants sensitive to abiotic stress display different ABA levels and metabolism during germination. Seed Science Research. 18: 249-258.
5. Sgroy, Verónica; Cassán, Fabricio ; Masciarelli, Oscar; Del Papa, Florencia , Lagares, Antonio, Luna, Virginia. 2009. Isolation and characterization of endophytic plant growth-promoting (PGPB) or stress homeostasis-regulating (PSHB) bacteria associated to the halophyte Prosopis strombulifera. Appl Microbiol Biotechnol. 85 (2): 371-382.
6. Fabricio Cassán, Oscar Masciarelli and Virginia Luna. 2009. Plant Growth Promoting Rhizobacteria Assays. In: Plant Bioassays. (Eds. S Narwal, A. Sampietro, N. Catalán and M. Vattuone, B. Polytika). Chapter 9, Pp 157-185. Editorial: Studium Press, LLC-USA.
7. Analía Llanes, Mariana Reginato, Gabriela Palacio and Virginia Luna. 2011. Biochemical indicators of salinity tolerance in the halophyte Prosopis strombulifera are differentially affected by NaCl and Na2SO4. . En: Urbanisation, Land Use, Land

Degradation and Environment. Chapter 22, pp. 344-355. Eds. Öztürk M., Mermut A., Çelik A. Daya Publishing House, Delhi-110002. India.

8. M. Reginato, V. Sgroy, A. Llanes, F. Cassán and V. Luna. 2011. The american halophyte Prosopis strombulifera, a new potential tool to confer salt tolerance to crops. In: Crop Production for Agricultural Improvement, Book Section: Breeding for Salinity and Drought Tolerance. NAM S&T Centre Press. Accepted, in press.
9. Mariana Reginato, Laura Sosa, Analia Llanes, Eugenio Hampp, Nelio Vetorazzi, Herminda Reinoso, and Virginia Luna. 2011. Ionic interactions within the major salts in the soil determine differential physiological responses in the halophytic legume Prosopis strombulifera . Botany (ex Canadian Journal). Under revision.
10. Mariana A. Reginato, Guillermina I. Abdala, Oscar A. Ruiz, Elsa Moschetti, Virginia Luna. 2011. Jasmonate and free polyamine levels in roots and leaves of the halophyte Prosopis strombulifera are differentially affected by Na2SO4 and NaCl. Biologia. Under revision.
11. Fabricio Cassán, Diego Perrig, Verónica Sgroy, Virginia Luna. 2011. Basic and Technological Aspects of Phytohormone Production by Microorganisms: Azospirillum sp. as a Model of Plant Growth Promoting Rhizobacteria. In: Plant growth and Health Promoting Bacteria: Plant Nutrient Management Vol. III.(D. K. Maheshwari, ed) SPRINGER In press.
12. Llanes A., Vidal S., Capdevielle F., Bonercarrere V., and Luna V. 2011. Genetic diversity in natural populations of halophyte legume Prosopis strombulifera revealed by AFLP fingerprinting. Boletín de la Sociedad Argentina de Botánica. En revisión.
13. Analía Llanes, Sabina Vidal, Oscar Masciarelli, Roxana Ordoñez, María Inés Isla, and Virginia Luna. 2011. Abscisic acid (ABA) signalling and salt-induced gene expression are differentially affected by NaCl and Na2SO4 in the halophyte Prosopis strombulifera: a key role for ABA-glucose ester. Journal of Plant Growth Regulation. En revision.
14. Furlan A., Llanes A., Luna V., Castro S. Abscisic acid and hydrogen peroxide trigger the physiological and biochemical responses of nodulated peanut to drought stress. 2011. Journal of Plant Growth Regulation. En revision.

In: Chemistry Researcher Biographical Sketches
Editors: E. S. Huseynov and V. Babayev
ISBN: 978-1-62100-771-5

Chapter 51

BIOGRAPHICAL SKETCH

Name:	Title:	Date of Birth:
Randall D Maples	Oklahoma State University Department of Chemistry	Durant

EDUCATION:

Institution and Location	Degree	Year Conferred	Scientific Field
Chemistry Southwestern Oklahoma State University.	B.S.	2007	
Chemistry Oklahoma State University	PhD	2012	

CONTACT POINTS:

23 Summit Circle, Stillwater, OK 74075

RESEARCH AND PROFESSIONAL EXPERIENCE:

Nanotechnology, Chemical weapons detection, Toxicology
Professional Appointments: American Chemical Society member

Publications during last three years

Binding Optimization through Coordination Chemistry: CXCR4 Chemokine Receptor Antagonists from Ultrarigid Metal Complexes

- Abid Khan, Gary Nicholson, John Greenman, Leigh Madden, Graeme McRobbie, Christophe Pannecouque, Erik De Clercq, Robert Ullom, Danny L. Maples, Randall D. Maples, Jon D. Silversides, Timothy J. Hubin and Stephen J. Archibald
- J. Am. Chem. Soc., 2009, 131 (10), pp 3416–3417

Synthesis and characterization of the chromium(III) complexes of ethylene

- Cross-bridged cyclam and cyclen ligands
- Danny L. Maples, Randall D. Maples, Wesley A. Hoffert, Trenton H. Parsell, Allan van Asselt,
- Jon D. Silversides, Stephen J. Archibald, Timothy J. Hubin
- Inorganica Chimica Acta 362 (2009) 2084–2088

In: Chemistry Researcher Biographical Sketches
Editors: E. S. Huseynov and V. Babayev

ISBN: 978-1-62100-771-5

Chapter 52

BIOGRAPHICAL SKETCH

Name:	Title:	Date of Birth:
María de los Ángeles Martín Santos	Chemical Engineering Department. University of Cordoba (Spain)	2th June, 1972

EDUCATION:

Institution and Location	Degree	Year Conferred	Scientific Field
Chemical Engineering	PhD	December 2001	

CONTACT POINTS:

Campus Universitario de Rabanales, Carretera Madrid-Cadiz, km 396. Edificio Marie Curie (C-3), planta baja. CP 14071, Cordoba, Spain. Telf.: +34 609768947. Fax: +34 957218625. Email: iq2masam@uco.es

RESEARCH AND PROFESSIONAL EXPERIENCE:

Biodiesel manufacturing from used cooking oils by alkaline transesterification in open and closed systems. Wastewater treatment:phisical, chemical and biological treatments. (Advanced oxidation processes for wastewater treatment, Physical-chemical treatments of oily wastewater). Anaerobic digestion of wastewater, wastes and organic industrial wastes.(Glyceriol, orange peel waste, wastewater derived from the pressing of orange peel generated in the orange juice production, vinasse and maize as energy crop). Inhibition of termophilic anaerobic digestion. Kinetic modeling of biological and chemical processes. Squalene purification from olive oil and shark liver. Composting of urban waste.

AWARDS AND HONORS:

Five awards on highly polluting wastes management

Professional Appointments:

Lecturer

Publications during last three years

1) Siles, J.A., Martín, M.A., Martín, A., Raposo, F., Borja, R., 2007. Anaerobic digestion of wastewater derived from the pressing of orange peel generated in the orange juice production. J. Agr. Food Chem., 55(6), 1905-1914.
2) Siles, J.A., Chica, A., Borja, R., Martín, M.A., 2007. Estudio y modelización de la digestión anaerobia de las aguas residuales del prensado de la cáscara de naranja. Tecnología del Agua, 27(291), 44-60.
3) Berrios, M., Siles, J.A, Martín, M.A., Martín, A., 2007. A kinetic study of the esterification of free fatty acids (FFA) in sunflower oil. Fuel, 86, 2383-2388.
4) Siles, J.A., Martín, M.A., Chica, A.F., Borja, R., 2008. Kinetic modelling of the anaerobic digestion of wastewater derived from the pressing of orange rind produced in orange juice manufacturing. Chem. Eng. J., 140, 145-156.
5) Chica, A.F., Martín, M.A., Siles, J.A., Dios, M., 2008. La medida de la actividad respiratoria durante el compostaje de residuos sólidos municipales. Reciclado y plásticos, 29, 7-8.
6) Siles, J.A., Martín, M. A., Chica, A.F., Martín, A., 2009. Anaerobic digestion of glycerol derived from biodiesel manufacturing. Bioresource Technol., 100(23), 5609-5615.
7) Martín, M.A., De la Rubia, M.A., Martín, A., Borja, R., Montalvo, S., Sánchez, E. 2010. Kinetic evaluation of the psychrophylic anaerobic digestion of synthetic domestic sewage using an upflow filter. Bioresource Technology 101 (1), pp. 131-137
8) Siles, J.A., Martín, M.A., Chica, A.F., Martín, A., 2010. Anaerobic co-digestion of glycerol and wastewater derived from biodiesel manufacturing. Bioresource Technol., 101, 6315-6321.
9) Martín, M.A., Siles, J.A., Chica, A.F., Martín, A., 2010. Modelling the anaerobic digestion of wastewater derived from the pressing of orange peel produced in orange juice manufacturing. Bioresource Technol., 101, 3909?3916.
10) Siles, J.A., Brekelmans, J., Martín, M.A., Chica, A.F., Martín, A. 2010. Impact of ammonia and sulphate concentration on thermophilic anaerobic digestion. Bioresource Technol., 101, 9040-9048.
11) Martín, M.A., Siles, J.A., Chica, A.F., Martín, A. 2010. Biomethanization of orange peel waste. Bioresource Technol., 101, 8993-8999.
12) J.A. Siles, I. García, A. Martín, M.A. Martín, 2011. Integrated ozonation and biomethanization treatments of vinasse derived from ethanol manufacturing. Journal

Hazardous and Materials. In Press, Accepted Manuscript, Available online 1 February 2011 (DOI: 10.1016/j.jhazmat.2011.01.096)

Patents:

1) Aguado, M., Chica, A., Berrios, M., Siles, J.A., Martín, M.A., Martín, A. A method to obtain squalene from animal origin. Nº P200700462. Date: 22nd of Febrary, 2007.
2) Aguado, M., Chica, A., Berrios, M., Siles, J.A., Martín, M.A., Martín, A. A method to obtain squalene from vegetable origin. Nº P200701369. Date: 18th of May, 2007.

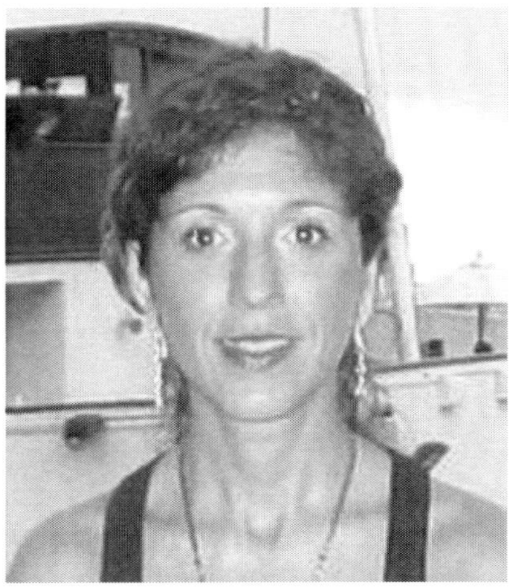

In: Chemistry Researcher Biographical Sketches
Editors: E. S. Huseynov and V. Babayev
ISBN: 978-1-62100-771-5

Chapter 53

BIOGRAPHICAL SKETCH

Name:	Title:	Date of Birth:
Jyri-Pekka Mikkola	Prof, Dr.Sc. (Chem.Eng.)	08.03.1966

EDUCATION:

Institution and Location	Degree	Year Conferred	Scientific Field
Technical Chemistry, Chemical-Biological Center, Umeå University, SE-901 87 Umeå, Sweden			
Industrial Chemistry and Reaction Engineering, Process Chemistry Centre, Åbo Akademi University, FI-20500 Åbo-Turku, Finland			

CONTACT POINTS:

Olof Palmes gata 29 SE-90323 Umeå, Sweden; Kiekkotie 6
FI-20360 Åbo-Turku, Finland

RESEARCH AND PROFESSIONAL EXPERIENCE:

Various ionic liquids technologies; chemical kinetics and reactor modeling; heterogeneous catalysis with special emphasize on structured catalysts; process and reaction intensification; nanocatalysis by gold; green chemistry from renewable sources: carbohydrate chemistry, catalysis of nature-derived products and novel materials.

AWARDS AND HONORS:

- 2006 - The Incentive Award by the Academy of Finland - the award for new, innovative research.
- 2009 - Umeå University Young Researcher Award year and
- 2010 - One of his Ph.D. students, Dr.Sc. Pasi Virtanen was elected as the winner of the Finnish Catalyst Society Award for the best doctoral thesis in Finland.

Professional Appointments:

Umeå University, Sweden & Åbo Akademi University, Finland, Professor in Technical Chemistry, Sustainable Development (appointed 2008) Academy of Finland (Åbo Akademi), Academy Research Fellow, 2004 – 2009 Åbo Akademi, Lab. of Ind. Chem., Researcher, Senior Research Associate, Postdoc. Researcher and Research Associate (lecturer) positions 1996 ? 2004

Publications during last three years

- Toukoniitty, B.; Mikkola, J.-P.; Murzin, D. Yu.; Salmi, T. Utilization of electromagnetic and acoustic irradiation in enhancing heterogeneous catalytic reactions. Trends in Chemical Engineering (2008), 11 1-37.
- S. Leveneur, D. Yu. Murzin, T. Salmi, J.-P. Mikkola, N. Kumar, K. Eränen, L. Estel, Synthesis of Peroxypropionic Acid from Propionic Acid and Hydrogen Peroxide over Heterogeneous Catalysts, Chem.Eng.J., 2008 doi:10.1016/j.cej.2008.11.045
- Salmi, Tapio; Kuusisto, Jyrki; Warna, Johan; Mikkola, Jyri-Pekka.Detailed kinetic analysis reveals the true reaction path: catalytic hydrogenation, hydrolysis and isomerization of lactose. Chemical Industries (Boca Raton, FL, United States) (2009), 123(Catalysis of Organic Reactions), 103-115.
- P. Mäki-Arvela, S. Sahin, N. Kumar, J.-P. Mikkola, K. Eränen, T. Salmi, D. Yu. Murzin, Utilization of cascade chemo-bio catalysis for synthesis of R-1-phenylethyl acetate, Reaction Kinetics and Catalysis Letters, 2008, 94, 281-288.
- Salmi, T., Wärnå, J, Mikkola, J.-P. CHEMICAL REACTION ENGINEERING AND REACTOR TECHNOLOGY, a text book in chemical kinetics and reactor modeling 2010 (Taylor & Francis), ISBN10: 1420092685
- Eta, V., Mäki-Arvela, P., Kordas, K., Murzin, D. Yu., Salmi, T., Mikkola, J.-P., The Activity of Zirconia and other Metal oxides in the Synthesis of Dimethyl Carbonate from Methanol and Carbon Dioxide, Book chapter in Progress in Heterogeneous Catalysis, Ed. Dieter L. Marmaduke, Nova Science Publishers, New York, 2008 ISBN: 978-1-60456-981-0
- Tapio Salmi, Dmitry Murzin, Päivi Mäki-Arvela, Johan Wärnå, Kari Eränen, Narendra Kumar, Jyri-Pekka Mikkola, Catalytic processing of biomass into chemicals, Ed. Cybulski , Wiley (submitted)

- Tokarev, A., Murzin, D. Yu., Mikkola, J.-P., Applied Electrochemical Methods for Heterogeneous Catalysis, Lambert Academic Publishing AG & Co. Ltd., Saarbrücken, Deutschland, 2009, ISBN: 978-3-8383-0975-0
- J.R. Hernandez Carucci, J. Sosa, K. Arve,, H. Karhu, J.-P. Mikkola, K. Eränen, T. Salmi, D.Yu. Murzin, Preparation and characterization of coated microchannels for the selective catalytic reduction of NOx, Catalysis: Principles, Types and Applications, Editor: Minsuh Song, pp., Nova Science Publishers, New York, 2011, ISBN 978-1-61209-654-4.
- Book chapter: ?Heterogeneous catalysis in biomass to chemicals and fuels?, Ed. Dr. D.Kubicka; Ionic liquids as catalytic medium for biomass transformations, E. Leino, V. Eta, S. Hyvärinen, E.Privalova, E. Salminen, J. Gräsvik, P.Virtanen, P. Mäki-Arvela, J.-P. Mikkola, 2011 (submitted)
- Päivi Mäki-Arvela, Serap Sahin, Narendra Kumar, Jyri-Pekka Mikkola, Kari Eränen, Tapio Salmi and Dmitry Yu. Murzin, One-pot utilization of heterogeneous and enzymatic catalysis: Synthesis of R-1-phenylethyl acetate from acetophenone, Cat.Today., 2008, 140 (2009) 70?73; doi:10.1016/j.cattod.2008.07.037
- D. Yu. Murzin, E. V. Murzina, A. V. Tokarev, and J.-P. Mikkola, Lactose Oxidation Kinetics with Oxygen in Catalyst?Solution?Gas Three?Phase System with Simultaneous Electrical Potential Measurement of Supported Gold Catalyst, Russian Journal of Electrochemistry, 2009, Vol. 45, No. 9, pp. 1017?1026
- Janek Reinik, Ivo Heinmaa, Jyri-Pekka Mikkola and Uuve Kirso, Synthesis and characterization of calcium?alumino-silicate hydrates from oil shale ash ? Towards industrial applications, Fuel 2007, doi:10.1016/j.fuel.2007.11.008
- T. Salmi, D.Yu. Murzin, P.Mäki-Arvela, J. Wärnå, K. Eränen, J.-P.Mikkola, A. Denecheau, K. Alho, Enhancement of three-phase kinetic studies of complex reaction systems by semibatch reactor technology, Annales Universitatis Mariae Curie-Sklodowska, Chemia (Lublin), 2008, 63, 257-270.
- Päivi Mäki-Arvela; Jyrki Kuusisto; Eloisa M Sevilla; Irina Simakova; J.-P. Mikkola; Jukka Myllyoja; Tapio Salmi; Dmitry Murzin, Catalytic hydrogenation of linoleic acid to stearic acid over different Pd and Ru supported catalysts, Applied Catalysis A: General, 2008 345, 201-212.
- Toukoniitty, B., Mikkola, J.-P., Murzin, D.Yu. and Salmi, T., Utilization of Electromagnetic and Acoustic Irradiation in Enhancing Heterogeneous Catalytic Reactions, a Review, Trends in Chem. Eng., vol.11, 2008,1-37.
- Toukoniitty, B.; Wärna, J.; Mikkola, J.-P.; Helle, M.; Saxen, H.; Murzin, D. Yu.; Salmi, T. Modelling of enantioselective and racemic hydrogenation of ethyl pyruvate on a Pt/Al2O3 catalyst in the presence of microwave irradiation. Chemical Engineering and Processing (2009), 48(4), 837-845.
- Olga A. Simakova, Anne-Riikka Leino, Betiana Campo, Päivi Mäki-Arvela, Krisztián Kordás, Jyri-Pekka Mikkola, and Dmitry Yu. Murzin, Linoleic acid isomerization over mesoporous carbon supported gold catalysts; Catalysis Today,150 (2010) 32?36 doi:10.1016/j.cattod.2009.07.065
- Olatunde Jogunola, Tapio Salmi, Johan Wärnå, Jyri-Pekka Mikkola and Esko Tirronen, kinetics of methyl formiate hydrolysis in the absence and presence of complexing agents, Chem.Eng.Sci 2009 (submitted)

- B. Toukoniitty, J. Wärnå, J.-P.Mikkola, T. Salmi, D.Yu. Murzin, M. Helle, H. Saxen, Modelling of enantioselective and racemic hydrogenation of ethyl pyruvate on a Pt/Al2O3 catalyst in the presence of microwave irradiation. Chemical Engineering and Processing: ProcessIntensification 2009 (accepted).
- Virtanen, Pasi; Karhu, Hannu; Toth, Geza; Kordas, Krisztian; Mikkola, Jyri-Pekka. Towards one-pot synthesis of menthols from citral: Modifying Supported Ionic Liquid Catalysts (SILCAs) with Lewis and Bronsted acids. Journal of Catalysis (2009), 263(2), 209-219. doi:10.1016/j.jcat.2009.02.005
- Virtanen, Pasi; Mikkola, Jyri-Pekka; Toukoniitty, Esa; Karhu, Hannu; Kordas, Krisztian; Eraenen, Kari; Waerna, Johan; Salmi, Tapio. Supported ionic liquid catalysts - From batch to continuous operation in preparation of fine chemicals. Catalysis Today (2009), 147(Suppl.), S144-S148.
- Mikkola, J.-P., Toukoniitty, B., Salmi, T. and Rochas, J.-F., Novel Laboratory-scale Multi-purpose Microwave Reactor Setup (MMRS) for Chemical Engineering Studies, Chem.Eng.J. (submitted)
- A.V.Tokarev, A. Kirilin, E. V. Murzina, N.V. Kulkova, J.-P. Mikkola, D.Yu. Murzin, Catalyst deactivation in aqueous phase reforming of sorbitol, Catalysis Today Special Issue 2009 (submitted)
- Mäki-Arvela, Päivi; Sahin, Serap; Kumar, Narendra; Mikkola, Jyri-Pekka; Eränen, Kari; Salmi, Tapio; Murzin, Dmitry Yu. One-pot utilization of heterogeneous and enzymatic catalysis: Synthesis of R-1-phenylethyl acetate from acetophenone. Catalysis Today (2009), 140(1-2), 70-73
- Maki-Arvela, Paivi; Sahin, Serap; Kumar, Narendra; Mikkola, Jyri-Pekka; Eranen, Kari; Salmi, Tapio; Murzin, Dmitry Yu. Utilization of cascade chemo-bio catalysis for the synthesis of R-1-phenylethyl acetate. Reaction Kinetics and Catalysis Letters (2008), 94(2), 281-288
- Jarmo Kukkola, Aatto Rautio, Giovanni Sala, Flavio Pino, Geza Toth, Anne-Riikka Leino, Jani Mäklin, Heli Jantunen, Antti Uusimäki, Krisztian Kordas, Eduardo Gracia, Mauricio Terrones, Andrey Shchukarev and Jyri-Pekka Mikkola, Electrical transport through single-wall carbon nanotube-anodic aluminum oxide-aluminum heterostructures, Nanotechnology 21 (2010) 035707, doi:10.1088/0957-4484/21/3/035707
- R.Pezoa, V.Cortinez, S.Hyvärinen, M.Reunanen, J.Hemming, M.E.Lienqueo, O.Salazar, R.Carmona, A.Garcia, D.Yu.Murzin, J.-P. Mikkola, The use of ionic liquids in the pretreatment of forest and agricultural residues for the production of bioethanol, Cellulose Chemistry and Technology 44 (1-3) 2010
- Pasi Virtanen, Tapio O. Salmi, Jyri-Pekka Mikkola, Supported Ionic Liquid Catalysts (SILCA) for Preparation of Organic Chemicals, ORCS 2010, Topics in Catalysis, vol.35, 15, 2010, 1096
- Privalova, E., Mäki-Arvela, P., Murzin, D. Yu., Mikkola, J.-P., Capturing CO2: Conventional versus non-conventional technologies for biogas plants, Biomass and Bioenergy (submitted)
- V.Cortinez, S. Hyvärinen, M. Reunanen, J. Hemming, M.E. Lienqueo, D. Yu. Murzin, J.-P. Mikkola, Evaluation of ionic liquid mediated pretreatment for the enzymatic

hydrolysis of forest residues to obtain bioethanol, Cellulose Chemistry and Technology 2010 44 (1-3)

- S. Hyvärinen, P. Virtanen, D. Yu. Murzin, J.-P. Mikkola, Towards Ionic Liquid Fractionation of Lignocellulosics for fermentable sugars, Cellulose Chemistry and Technology 2010 44 (1-3)
- J.Reinik, I. Heinmaa, J.-P. Mikkola, K. Kordas, U. Kirso, Oil Shale, 2010, Vol. 27, No. 1, pp. 47-57 doi: 10.3176/oil.2010.1.06 (Estonian Academy Publishers)
- Pia Damlin, Jyri-Pekka Mikkola, Tapio Salmi, Characterization of hardwood-derived carboxymethylcellulose by high-pH anion-exchange chromatography using pulsed amperometric detection, Cellulose Chem. Technol., 44 (1-3), 65-69 (2010) (in press)
- Tapio Salmi, Pia Damlin, Jyri-Pekka Mikkola, Matias Kangas, Modelling and Experimental Verification of Cellulose Dissolution Kinetics, Chem.Eng.Sci, 2010 (accepted)
- P.Mäki-Arvela, I. Anugwom, P. Virtanen, R. Sjöholm, J.P. Mikkola, Dissolution of lignocellulosic materials and its constituents using ionic liquids?A review, Ind. Crops and Prod. 32 (2010) 175-201, doi:10.1016/j.indcrop.2010.04.005
- A. Kirilin , S. Sahin, A. Tokarev, P. Maki-Arvela, K. Kordas, A.-R. Leino, J.-P. Mikkola,, L. M. Kustov, T. Salmi, D. Yu. Murzin, Mechanistic investigations of the reaction network in chemo-bio catalyzed synthesis of R-1-phenylethylacetate, ? 374, Kinetics & Catalysis 2010 (accepted)
- Alexey V. Kirilin, Anton V. Tokarev, Elena V. Murzina, Leonid M. Kustov, Jyri-Pekka Mikkola and Dmitry Yu. Murzin, Reaction Products and Transformations of Intermediates in the Aqueous Phase Reforming of Sorbitol, ChemSusChem, 3, 708-718, 2010, doi: 10.1002/cssc.200900254
- Pia Hara, Jyri-Pekka Mikkola, Dmitry Yu. Murzin and Liisa Kanerva, Supported Ionic Liquids in Burkholderia cepacia Lipase-Catalyzed Asymmetric Acylation, J.Mol.Cat. B, Enzymatic, 2010 MOLCAB2112PII: S1381-1177(10) 00202-X DOI: 10.1016/j.molcatb.2010.07.018
- E. Leino, P. Mäki-Arvela, V. Eta, D. Yu. Murzin, T. Salmi, J.-P. Mikkola, Conventional synthesis methods of short-chain dialkylcarbonates and novel production technology via direct route from alcohol and waste CO2 ? A Review, Applied Catalysis A: General 383 (2010) 1?13, doi:10.1016/j.apcata.2010.05.046
- Matias Kangas, Tapio Salmi, Pia Damlin, Jyri-Pekka Mikkola, Modelling and Experimental Verification of Cellulose Substitution Kinetics, Chem.Eng.Sci, 2010 (in press)
- J.R. Hernandez Carucci, J. Sosa, K. Arve,, H. Karhu, J.-P. Mikkola, K. Eränen, T. Salmi, D.Yu. Murzin, Preparation and characterization of coated microchannels for the selective catalytic reduction of NOx, Catalysis: Principles, Types and Applications (Nova Publishers) (accepted).
- Murzin, Dmitry; Grenman, Henrik; Warna, Johan; Mikkola, J-P; Sifontes, Victor; Fardim, Pedro; Salmi, Tapio, Modelling the influence of wood anisotropy and internal diffusion on delignification kinetics, Ind.Eng.Chem.Res. 2010, Ind. Eng. Chem. Res. 2010, 49, 9703?9711, doi: 10.1021/ie101215a

- A.V. Tokarev, A.V. Kirilin, E.V. Murzina, K. Eränen, L.M. Kustov, D.Yu. Murzin, J.-P. Mikkola, The role of bio-ethanol in aqueous phase reforming to sustainable hydrogen, Int.J.Hydr.En., 35 (2010) 12642 12649 doi:10.1016/j.ijhydene.2010.07.118
- Valerie Eta, Päivi Mäki-Arvela, Anne-Riikka Leino, Krisztain Kordas, Tapio Salmi, Dmitry Yu. Murzin, Jyri-Pekka Mikkola, Synthesis of Dimethyl Carbonate from Methanol and Carbon Dioxide: Circumventing Thermodynamic Limitations, Ind. Eng. Chem. Res. 2010, 49, 9609?9617, doi: 10.1021/ie1012147
- S. Ajaikumar, J. Ahlkvist, W. Larsson, A. Shchukarev, K. Kordas, J.- P. Mikkola, Oxidation of ?-pinene over gold containing bimetallic nanoparticles supported on reducible TiO2 by DPU method, Appl.Cat. A:General, 2010, doi: 10.1016/j.apcata.2010.10.015 (in press)
- Mikkola, J.-P., Biorefining of Nordic Woody Lignocellulose – the Emerging Industry, International Conference and Fair on Education, Research & Employment in the Renewable Energy Sector- ICERE (2) 2010, Santiago, Chile, 11th- 12th November 2010 (published in ?.)
- Víctor A. Sifontes Herrera, Oluwamuyiwa Oladele, Krisztián Kordás, Kari Eränen, J. P. Mikkola Dmitry Yu. Murzin and Tapio Salmi, Sugar Hydrogenation over a Ru/C Catalyst, J.Chem.Tech.Biotech. 2011 DOI 10.1002/jctb.2565
- A. Kirilin, S. Sahin, A. Tokarev, P. Mäki Arvela, K. Kordas, A, R. Leino, A. Shchukarev, J-P. Mikkola, L. M. Kustov, T. Salmi, and D. Yu. Murzin, Mechanistic Investigations of the Reaction Network in Chem-Bio Catalyzed Synthesis of R-1-Phenylethyl Acetate, VIII INTERNATIONAL CONFERENCE ON MECHANISMS OF CATALYTIC REACTIONS, ISSN 002-1584, Kinetics and Catalysis, 2010, Vol. 51, No. 6, pp. 809?815. © Pleiades Publishing, Ltd., 2010.
- Ming-Chung Wu, Jussi Hiltunen, András Sápi, Anna Avila, William Larsson, Hseuh-Chung Liao, Mika Huuhtanen, Géza Tóth, Andrey Shchukarev, Ákos Kukovecz, Zoltán Kónya, Jyri-Pekka Mikkola, Riitta Keiski, Wei-Fang Su, Heli Jantunen, and Krisztián Kordás, Nitrogen-doped TiO2 Anatase Nanofibers Decorated with Noble Metal Nanoparticles for High-efficiency Photocatalytic Hydrogen Generation, Int.J.Hydr.En. 2011 (submitted)
- A. Kirilin, P. Mäki-Arvela, K. Kordas, A.-R. Leino, A. Shchukarev, D. Boström, J.-P. Mikkola, L. M. Kustov, T. O. Salmi, D. Yu. Murzin, Chemo-bio catalyzed synthesis of R-1-phenyethyl acetate over bimetallic PdZn catalysts, lipase and Ru/Al2O3, Part I, 2011, 52, 1, 73?77.
- A. Kirilin, P. Mäki-Arvela, K. Kordas, A.-R. Leino, A. Shchukarev, D. Boström, J.-P. Mikkola, L. M. Kustov, T. O. Salmi, D. Yu. Murzin, Chemo-bio catalyzed synthesis of R-1-phenyethyl acetate over bimetallic PdZn catalysts, lipase and Ru/Al2O3, Part II, 2011 52, ? 1, 8?82
- V. Kontturi, S. Hyvärinen, A. García, D. Yu. Murzin, J.-P. Mikkola, K.-E. Peiponen, Simultaneous detection of absorption spectrum and refractive index ratio with a spectrophotometer: monitoring contaminants in bioethanol, Measurement Science and Technology 2011 (accepted, in press)

In: Chemistry Researcher Biographical Sketches
Editors: E. S. Huseynov and V. Babayev

ISBN: 978-1-62100-771-5

Chapter 54

BIOGRAPHICAL SKETCH

Name:	Title:	Date of Birth:
Steven J. Milne	Dr	22 April 1957

EDUCATION:

Institution and Location	Degree	Year Conferred	Scientific Field
	PhD, BSc (1st class honours)		

CONTACT POINTS:

E-mail: s.j.milne@leeds.ac.uk University of Leeds, UK

RESEARCH AND PROFESSIONAL EXPERIENCE:

Reader in Ceramic Science, Inst. For Materials research, University of Leeds (Academic position since 1985)

AWARDS AND HONORS:

Series of invited talks and invited papers in special issues of journals- latest in IEEE TUFCC (in review).

Professional Appointments:

External PhD examiner

Publications during last three years

1. J S Earl, D J Wood, S J Milne. “Hydroxyapatite nanoparticle synthesis and in-vitro study of dentinal tubule occlusion”. NSTI Nanotechnology Conference and Trade Show, Boston, MA, United States. 1, 407-10 (2006).
2. P. Bomlai, N Sirikulrat, A P Brown, E Condliffe, S J Milne. Compositional analysis and electrical properties. J. Mat. Sci. 42, 2175-80 (2007).
3. P Bomlai, S J Milne ”Properties of barium-strontium titanate PTCR ceramics sintered on different powder beds”. J. Mat. Sci. 42, 6803-8 (2007).
4. P Bomlai, P Wichianrat, S Muensit, S J Milne. “Effect of calcination conditions and excess alkali carbonate on the phase formation and particle morphology of Na0.5K0.5NbO3 powders”. J. Amer. Cer. Soc. 90, 1650-5 (2007).
5. C Puchmark, S Jiansirisomboon, G Rujijanagul, T P Comyn, J Y He, S J Milne. “Properties of lead zirconate-alumina 'nanocomposites”. Materials Research Bulletin; 42, 1269-77 (2007).
6. T A Skidmore, S J Milne. “Phase development during mixed-oxide processing of a [Na0.5K0.5NbO3](1-x)[LiTaO3](x) powder”. J. Mat. Res. 22 2265-72 (2007).
7. J Y He, T P Comyn, T A Skidmore, S J Milne, A J Bell. “Molten-salt synthesis of bismuth titanate and fabrication of TbTiO3-based textured ceramics”. 15th IEEE International Symposium of Applications of Ferroelectrics, IEEE Int Ferro (2007).
8. A J Scott, W Nimmo, R Scott, A P Brown, S J Milne. “Effects of aerosol heating rate on the properties of aggregates of lead zirconate titanate nanoparticles produced by spray pyrolysis”. J. Mater. Sci., 43, 6353-6358 (2008).
9. P Bomlai, S Sukprasert, S J Milne. “Reaction-sintering of lead-free piezoceramic compositions: (0.95-x)Na0.5K0.5Nb03.0.05LiTaO(3)-xLiSbO(3)”. J. Mater. Sci., 43, 6116-6121 (2008.
10. A Chowdhury, P R Thompson, S J Milne. “TGA-FFIR study of a lead zirconate titanate gel made from a triol-based sol-gel system”. Thermochim Acta 475, 59-64 (2008).
11. T P Comyn, T Chakrabory, R E Miles, S J Milne. ”Characterization of laser-transferred bismuth ferrite lead titanate ferroelectric thick films”. Appl. Phys. Lett., 93 Art No. 052909 (2008).
12. T Bongkarn, G Rujijanagul, S J Milne. “Antiferroelectric-ferroelectric phase transitions in Pb1-xBaxZrO3 ceramics: Effect of PbO content”. Appl. Phys. Lett., 92 Art No. 092905 (2008).
13. P Bomlai, P Sinsap, S Muensit, S J Milne. “Effect of MnO on the phase development, microstrues, and dielectric properties of 0.95Na(0.5)K(0.5)NbO(3)-0.05LiTaO(3)”. J. Am. Ceram. Soc, 91 624-627 (2008).
14. A Chowdhury, M A Khan, C James, S J Milne. “Densification and enhanced polarization in lead zirconate titanate sol-gel thin films” Mater. Chem. Phys. 113, 135-139 (2009).
15. P. Bomlai, S. Muensit, S J Milne. ‘ Phase transition behaviour and dielectric properties of $Na_{0.5}K_{0.5}NbO_3$-$LiTaO_3$ -$LiSbO_3$ lead-free piezoceramic’ Science Asia 35, 261-267, (2009)

16. T A Skidmore, T P Comyn, S J Milne. 'Temperature stability of ([$Na_{0.5}K_{0.5}NbO_3$](0.93)- [$LiTaO_3$](0.07)) lead-free piezoelectric ceramics'. Appl. Phys. Lett 94(22) Art. No. 222902 01. (2009)
17. A Chowdhury, S O'Callaghan, T A Skidmore, C James, S J Milne. 'Nanopowders of $Na_{0.5}K_{0.5}NbO_3$ Prepared by the Pechini Method'. J. Am. Ceram. Soc., 92, 758-761 (2009)
18. A Chowdhury, M A Khan, C James, S J Milne. 'Densification and enhanced polarisation in lead zirconate titanate sol-gel thin films'. Mater. Chem. Phys. 113, 135-139 (2009)
19. T Chakraborty , B M Xu, Q Zhang, A J Bell, X Bo, A Chowdhury, C James, C Puchmark , J Harrington, M Khan, R E Miles, W Z Xiong, S J Milne. 'Laser Transfer Processing and the Integration of Ferroelectric Films,' 9th European Conference on Applications of Polar Dielectrics (ECAPD-9), Rome, Italy, 25 Aug 2008 - 29 Aug 2008. Integrated Ferroelectrics. 106, 40-48, (2009)
20. J S Earl, D J Wood, S J Milne. 'Nanoparticles for Dentine Tubule Infiltration: An In Vitro Study' J Nanosci. Nanotechno. 9, xxxx (2009)
21. C James, T Chakraborty, A Brown, T Comyn, R Dorey, J Harrington, A J Laister, R E Miles, C Puchmark, B Xu, R E Miles, W Z Xiong, S J Milne. 'Laser transfer processing for the integration of thin and thick film ferroelectrics' J. Mater. Sci. 44 xxxx (2009) (Invited)
22. A Chowdhury, J Bould, Y F Zhang, C James, S J Milne. 'Nano-powders of $Na_{0.5}K_{0.5}NbO_3$ made by a sol-gel method'. J. Nanoparticle. Res., 12(1) xxxx (2010)
23. T A Skidmore, T P Comyn, S J Milne. 'Dielectric and Piezoelectric Properties in the System (1- x)[(Na0.5K0.5NbO3)(0.93)-(LiTaO3)(0.07)]-x[BiScO3]' J. Am. Ceram. Soc., 93(3) xxxx (2010)

In: Chemistry Researcher Biographical Sketches
Editors: E. S. Huseynov and V. Babayev
ISBN: 978-1-62100-771-5

Chapter 55

BIOGRAPHICAL SKETCH

Name:	Title:	Date of Birth:
Vinita Mishra	MBBS, MD, MPhil, FRCPath	31st March, 1965

EDUCATION:

Institution and Location	Degree	Year Conferred	Scientific Field
Government Medical College Punjabi University, Patiala, India.	MBBS,	1983-1987:	
Internship, Government Medical College Punjabi University, Patiala, India		1988:	
Government Medical College Punjabi University, Patiala, India.	MD	1992-1995:	
Clinical Chemistry & Metabolic Medicine, Royal Liverpool & Broadgreen University NHS Trust, Liverpool, UK.	Senior House Officer	1995-1996:	
Lecturer, Department of Biochemistry, Government Medical College, Chandigarh, India.		1996-2000	
Clinical Research Fellow (MPhil), Clinical Chemistry & Metabolic Medicine, Royal Liverpool & Broadgreen University NHS Trust, Liverpool, UK.		2000-2002:	

Institution and Location	Degree	Year Conferred	Scientific Field
Specialist Registrar (FRCPath), Clinical Chemistry & Metabolic Medicine, Royal Liverpool & Broadgreen University NHS Trust, Liverpool, UK.		2002-2007	

CONTACT POINTS:

2, Coral Ridge, Wirral, Liverpool, CH43 7XE, UK

RESEARCH AND PROFESSIONAL EXPERIENCE:

- Estimation of total antioxidant capacity by enhanced chemiluminesence. Department of Clinical Pharmacology, Western General Hospital, Edinburgh, UK (2000).
- Estimation of lipid peroxidation markers such as Malondialdehyde and F2 Isoprostanes in critically ill patients. Clinical Chemistry, RLUH, UK (2001).
- PRISM trial on Paget's patients. Clinical Chemistry, RLUH, UK (2003).
- Cell culture and its response towards free radical generation on antioxidant. University of Louisiana, Department of Physiology, USA (2003).
- Hyperoxia study. Dept of Clinical Pharmacology, Western General Hospital, Edinburgh (2002).
- Selenium supplementation in critically ill patient. Clinical Chemistry, Intensive therapy Unit, Department of Immunology, RLUH, UK (2004).
- Estimation of glomerular filtration rate in-patients with Type 2 diabetes. Department of Medicine & Clinical Chemistry, Aintree University Hospital, UK (2006).
- The effect of gut hormone, glucagon-like peptide-1, on bone metabolism in normal and osteoporotic subjects. Clinical Chemistry, RLUH UK (2006).
- Incretin effect in South Asians. Clinical Chemistry, RLUH, UK (2006).

Professional Experience:

- Clinical Experience: Bone, Lipid, and renal stone metabolic clinics. Managing the hospitalized patients on nutritional support.
- Laboratory experience: managing biochemical abnormalities In the hospitalized patients.

AWARDS AND HONORS:

PRIZES

1986 Bronze medal in Pharmacology & Forensic Medicine, Government Medical College, Patiala, India

2002 First prize in the North West Association for Clinical Biochemists (ACB) meeting,Liverpool (Ian Ward paper) presentation “Correlation of markers of oxidative damage and antioxidant defense with clinical outcome in critically ill patients.”

AWARDS/HONOURS

May-June 2000 - CP Stewart Fellowship by the ACB, UK to work on the Total Antioxidant estimation by chemiluminescence’s method in the laboratory of Dr Simon Maxwell, Department of Pharmacology, Edinburgh University, UK,

August 2003 - Visiting fellowship at Louisiana State University Medical Center, School of Medicine, New Orleans, USA.

July 2005 -Associate Fellow by the NACB (National Academy of Clinical Biochemistry) USA.

April 2011 Outstanding speaker award for annual AACC (American Association of Clinical Chemistry) meeting, July 2010

COMMITTEE POSITIONS

Jan 2006 - Secretary for Nutrition division of AACC.

Jan 2009 Chair-Elect for Nutrition division of AACC.

May 2009 - Advisor for NCEPOD (national confidential enquiry into patient outcome and death) parenteral nutrition study.

Sep 2009 - Elected member of South West England Regional Council of Royal College of Pathologists,UK.

Jan 2010 - Chair for Nutrition division of AACC.

April 2010 - Regional Royal College of Pathologists Coordinator for “National Pathology Week 2010” for South West England, UK.

March 2011 - Public Engagement Regional Co-ordinator for South West England,UK.

Professional Appointments:

2007-2011: Consultant Chemical Pathologist, Laboratory Medicine, Salisbury NHS Foundation Trust, Salisbury, UK.

2011(July): Consultant Chemical Pathologist, Clinical Chemistry & Metabolic Medicine, Royal Liverpool & Broadgreen University NHS Trust, Liverpool, UK.

Publications during last three years:

Mishra V, Baines M, Parry S, Mclaughlin J, Carson J, Wenstone R, Shenkin A. "Effect of Selenium supplementation on biochemical markers and outcome in critically ill patients" Clinical Nutrition 2007; 26: 41-50.

Mishra V. Oxidative stress and role of antioxidant supplementation in critical illness", Review article. Clinical Laboratory 2007; 53: 199-209.

Marks E, Mishra V."The wider health considerations for calcium and vitamin D therapy"(in press, Dec 2010,accepted inOsteoporosis Bulletin).

S Nair, V Mishra, K Hayden, P Lisboa, B Pandya, S Vinjamuri, K J Hardy, J P H Wilding."The 4-variable MDRD formula underestimates glomerular filtration rate in the obese type 2 diabetic subjects with chronic kidney disease" (in press, Dec 2010,accepted in Diabetologia).

In: Chemistry Researcher Biographical Sketches
Editors: E. S. Huseynov and V. Babayev
ISBN: 978-1-62100-771-5

Chapter 56

BIOGRAPHICAL SKETCH

Name:	Title:	Date of Birth:
Yuichi Morishita	Senior Research Scientist & Leader of the Magma-Hydrothermal Deposits Research Group, Geological Survey of Japan, National Institute of Advanced Industrial Science and Technology	July 5, 1954

EDUCATION:

Institution and Location	Degree	Year Conferred	Scientific Field
Doctor of Science: Geology, University of Tokyo, March 1987 (Thesis title: Oxygen and carbon isotopic studies on hypothermal, skarn-type and epithermal ore deposits in Japan)			

CONTACT POINTS:

Central 7, Higashi 1-1-1, Tsukuba 305-8567, Japan

RESEARCH AND PROFESSIONAL EXPERIENCE:

Leader of the Secondary Ion Mass Spectrometry Laboratory
Leader of the Stable Isotope Laboratory

AWARDS AND HONORS:

Article Award of the Society of Resource Geology

Professional Appointments:

Officer, Nuclear Safety Bureau, Science and Technology Agency
Visiting Research Fellow, Department of Geological Sciences, Brown University, USA
Senior Researcher, Ore Genesis Section, GSJ, AIST
Senior Researcher & Leader, Ore Genesis Section, GSJ, AIST
Senior Researcher & Leader, Micro-scale Isotope Geochemistry Group, GSJ, AIST

Publications during last three years

Morishita, Y. and Takeno, N. (2010) Nature of the ore-forming fluid at the Quaternary Noya gold deposit in Kyushu, Japan, Resource Geology, 60, 359-376.

Saito, G., Morishita, Y. and Shinohara, H. (2010) Magma plumbing system of the 2000 eruption of Miyakejima volcano, Japan, deduced from volatile and major component contents of olivine-hosted melt inclusions, Journal of Geophysical Research, 115, B11202, 1-29.

Horie, K., Tsutsumi, Y., Cho, M., Morishita, Y. and Hidaka, H. (2010) Crystallization of REE minerals and redistribution of U, Th, and REE at contact boundary between granite and gabbro during hydrothermal alteration, Physics and Chemistry of the Earth, 35, 284-291.

Hammond, N.Q. and Morishita, Y. (2009) Source of ore fluids at the Kalahari Goldridge deposit, Kraaipan greenstone belt, South Africa: evidence from Sr, C and O isotope signatures in carbonates, Geofluids, 9, 356-364.

Morishita, Y., and Nakano, (2008) Role of basement in epithermal deposits: the Kushikino and Hishikari gold deposits, southwestern Japan, Ore Geology Reviews, 34, 597-609.

Morishita, Y., Shimada, N, and Shimada, K. (2008) Invisible gold and arsenic in pyrite from the high-grade Hishikari gold deposit, Japan, Applied Surface Science, 255, 1451-1454.

Nagahara, H., Kita, N.T., Ozawa, K. and Morishita, Y. (2008) Condensation of major elements during chondrule formation and its implication to the origin of chondrules, Geochimica et Cosmochimica Acta, 72, 1442-1465.

Kurahashi, E., Kita, N.T., Nagahara, H. and Morishita, Y. (2008) 26Al–26Mg systematicss of chondrules in a primitive CO chondrite, Geochimica et Cosmochimica Acta, 72, 3865-3882.

In: Chemistry Researcher Biographical Sketches
Editors: E. S. Huseynov and V. Babayev
ISBN: 978-1-62100-771-5

Chapter 57

BIOGRAPHICAL SKETCH

Name:	Title:	Date of Birth:
Jane Morrell	Swedish University of Agricultural Sciences (SLU), Uppsala, Sweden	14th June 1958

EDUCATION:

Institution and Location	Degree	Year Conferred	Scientific Field
BVet Med, BSc(Hons), PhD, MBA, Diplomate of Royal College of Veterinary Surgeons			

CONTACT POINTS:

Divison of Reproduction, Department of Clinical Sciences,
Swedish University of Agricultural Sciences,
Box 7054, SE-75007 Uppsala, Sweden.

RESEARCH AND PROFESSIONAL EXPERIENCE:

Veterinary andrologist with approximately 25 years independent research experience. She received her PhD in sperm sexing by flow cytometry 1987. Her other major research projects include making transgenic animals; collaborative work on the identification and preliminary investigation of a peptide in seminal plasma, fertilization promoting peptide (FPP); developing methods for semen collection, insemination and sperm cryopreservation in New World primates (for which she was awarded the Diploma of Fellowship of the Royal College of Veterinary Surgeons); research and development of more than 35 density gradient products and related products for many species. She has an extensive publication list, although she was unable to publish the majority of her research undertaken during her recent employment in industry ("proprietary information"). Her research interests include spermatology, sperm cryopreservation, reproductive biotechnology, gamete interaction, conservation of endangered species and protection of the environment.

Publications during last three years:

Morrell, J.M., Johannisson, A., Dalin, A-M., Hammar, L., Sandebert, T, & Rodriguez-Martinez, H. (2008) Sperm morphology and chromatin integrity on Swedish warmblood stallions and their relationship to pregnancy rates. Acta Vet Scand. 50, 2-8.

Thys, M., Vandaele L, Morrell JM, Mestach J, Van Soom A, Hoogewijs M, Rodriguez-Martinez H. (2009) *In vitro* fertilising capacity of frozen-thawed bull spermatozoa selected by single-layer glycidoxypropyltrimethoxysilane-coated silica colloidal centrifugation. Reproduction in Domestic Animals 44, 390-394.

Morrell, J.M., Dalin, A-M & Rodriguez-Martinez, H. 2008 Prolongation of stallion sperm survival by centrifugation through coated silica colloids: a preliminary study. Animal Reproduction 5, 121-126.

Morrell, J.M., Dalin, A-M., & Rodriguez-Martinez, H. (2009) Comparison of density gradient and single layer centrifugation of stallion spermatozoa: yield, motility and survival. Equine Veterinary Journal 41, 53-58.

Morrell, J.M., Johannisson, A., Strutz, H., Dalin, A-M. & Rodriguez-Martinez, H. (2009) Colloidal centrifugation of stallion semen: changes in sperm motility, velocity and chromatin integrity during storage. Journal Equine Veterinary Science 29, 24-32.

Morrell, J.M., Johannisson, A., Dalin, A-M, Rodriguez-Martinez, H. (2009) Morphology and chromatin integrity of stallion spermatozoa prepared by density gradient and single layer centrifugation through silica colloids. Reproduction Domestic Animals 44, 512-517.

Macias GB, Morrell J.M, Ortega Ferrusola C., Fernandez-Gonzalez L., Tapia JA, Rodriguez-Martinez H, Pena FJ. Centrifugation on a single layer of colloid selects improved quality spermatozoa from frozen-thawed stallion semen. Reproduction Domestic Animals 44, 523-526.

Macias GB, Fernandez-Gonzalez L., Morrell J., Ortega Ferrusola C., Tapia JA, Rodriguez-Martinez H, Pena FJ. (2009) Single layer centrifugation through colloid positively modifies the sperm sub-population structure of frozen-thawed stallion spermatozoa. Anim Reprod Sci 114, 193-202.

Morrell JM & Rodriguez-Martinez H. (2009) Biomimetic techniques for improving sperm quality in animal breeding: a review. The Open Andrology Journal 1, 1-9.

Johannisson, A., Morrell, J.M., Thorén, J, Jonsson, M, Dalin, A-M, Rodriguez-Martinez, H. (2009) Colloidal centrifugation with Androcoll-E™ prolongs stallion sperm motility, viability and chromatin integrity. Anim Reprod Sci. 116, 119-128.

Tejerina F, Morrell J, Petterson J, Dalin A-M, Rodriguez-Martinez H. (2009) Routine assessment of motility of ejaculated stallion spermatozoa using a novel computer-assisted motility analyser (Qualisperm™). Animal Reproduction 6, 380-385.

Ortega Ferrusola C, Fernandez-Gonzalez L, Morrell JM, Salazar Sandoval, C, Marcias GB, Rodriguez-Martinez H, Tapia JA, Pena FJ. (2009) Lipid peroxidation assessed with

BODIPY-C11 increases after cryopreservation of stallion spermatozoa, is stallion dependent and related to apoptotic changes. Reproduction, 138, 55-63.

Morrell, JM, Saravia, F, Wallgren M, van Wienen, M & Rodriguez-Martinez, H. (2009) Selection of boar spermatozoa using centrifugation on a glycidoxypropyltrimethoxy-silane-coated silica colloid. Journal of Reproduction and Development 55, 547-552.

Morrell, J.M., Johannisson, A., Dalin, A-M, Rodriguez-Martinez, H. (2009) Single Layer Centrifugation with Androcoll™-E can be scaled-up to allow large volumes of stallion ejaculate to be processed easily. Theriogenology 72, 879-884.

Morrell, JM, Saravia, F, Wallgren M, van Wienen, M & Rodriguez-Martinez, H. (2009) Sperm survival following colloid centrifugation varies according to the part of the sperm-rich fraction used. Society for Reproduction and Fertility 66, 85-86.

Morrell, J.M., Johannisson, A., Juntilla L, Rytty K, Bäckgren L, Dalin, A-M, Rodriguez-Martinez, H. (2010) Stallion sperm viability, as measured by the Nucleocounter SP-100, is affected by extender and enhanced by Single Layer Centrifugation. Veterinary Medicine International doi:10.4061/2010/659862

Chatdarong K, Thuwanut P, Morrell JM (2010) Single-layer centrifugation through colloid selects improved quality of epididymal cat sperm Theriogenology 73, 1284-92.

Morrell JM, Wallgren M. (2010) Androcoll™-P-Large selects boar spermatozoa with good membrane integrity from the sperm-rich fraction of the ejaculate. Animal Reproduction 7, 16-20.

Morrell JM, Rodriguez-Martinez H, Johannisson A. (2010) Single Layer Centrifugation of stallion spermatozoa consistently selects the most robust spermatozoa from the rest of the ejaculate in a large sample size: data from 3 breeding seasons. Equine Veterinary Journal 42, 579-585.

Morrell JM, Rodriguez-Martinez H, Johannisson A. (2010) Single Layer Centrifugation of stallion spermatozoa improves sperm quality compared to sperm washing. Reproductive Biomedicine Online 21, 429-436.

Morrell JM, Rodriguez-Martinez H. (2010) Practical applications of sperm selection techniques as a tool for improving reproductive efficiency. Veterinary Medicine International doi:10.4061/2011/2984767

Bergqvist A-S, Johannisson A, Bäckgren L, Dalin A-M, Rodriguez-Martinez H, Morrell JM. (2011) Single layer centrifugation of stallion spermatozoa through Androcoll™-E does not adversely affect their capacitation-like status, as measured by CTC-staining. Reprod Dom Anim 46, 74-78.

Blomqvist G, Persson M, Wallgren M, Wallgren P, Morrell JM. (2011) Removal of Virus from Boar Semen Spiked with Porcine Circovirus Type 2. Anim Reprod Sci 126, 108-114.

Ortega Ferrusola C, Johannisson A, Peña FJ, Tapia JA, Rodriguez-Martinez H, Dalin AM and JM Morrell (2011) Effect of different extenders and seminal plasma on the susceptibility of equine spermatozoa to lipid peroxidation after Single layer centrifugation (SLC), through Androcoll™-E. JEVS 31, 411-426.

Morrell JM, Johannisson A, Rodriguez-Martinez H (2011) Effect of osmolarity and density of colloid formulations on the outcome of SLC-selection of stallion spermatozoa. ISRN Veterinary Science doi:10.5402/2011/128984.

Jane M. Morrell (2011). Artificial Insemination: Current and Future Trends, Artificial Insemination in Farm Animals, Milad Manafi (Ed.), ISBN: 978-953-307-312-5, InTech,

Available from: http://www.intechopen.com/articles/show/title/artificial-insemination-current-and-future-trends

Morrell JM, Mari G, Kutvölgyi G, Meurling S, Iacono E, Mislei B, Rodriguez-Martinez H. (2011) Spermatozoa from stallion ejaculates processed by Single Layer Centrifugation with Androcoll™-E are capable of fertilization after artificial insemination. Reprod Dom Anim. 46, 642-645. Doi:10.1111/j.1439-0531.2010.01721x.

Morrell JM, Wallgren M. (2011) Removal of bacteria from boar ejaculates by Single Layer Centrifugation can reduce the use of antibiotics in semen extenders. Animal Reproduction Science 123, 64-9.

Van Wienen M, Johannisson, A., Wallgren, M, Joyce Parlevliet, J., Morrell JM. Single Layer Centrifugation with Androcoll™-P can be scaled-up to process larger volumes of boar semen. ISRN Veterinary Science *In press*

Colleoni, S., Lagutina, I., Rodriguez-Martinez, H., Lazzari, G., Galli C., Morrell, JM. New techniques for selecting spermatozoa for equine assisted reproduction. JEVS *in press*

Kruse, R, Dutta PC, Morrell JM. Colloid centrifugation removes seminal plasma and cholesterol from boar spermatozoa. Reproduction, Fertility and Development, *In press*

Morrell JM, Macias Garcia B, Pena FJ, Johannisson A. (2011) Processing stored stallion semen doses by Single Layer Centrifugation. Theriogenology, *in press.*

Hoogewijs M, Morrell JM, Van Soom A, Govaere J, Johannisson A, Piepers P, De Schauwer C, de Kruif A, De Vliegher S. (2011) Sperm selection using single layer centrifugation prior to cryopreservation can increase post thaw sperm quality in stallions. Equine Vet Journal, *In Press*

Morrell JM, Mari G, Kutvölgyi G, Meurling S, Iacono E, Mislei B, Rodriguez-Martinez H. Spermatozoa from stallion ejaculates processed by Single Layer Centrifugation with Androcoll™-E are capable of fertilization after artificial insemination. Reprod Dom Anim. 10.1111/j.1439-0531.2010.01721.x

Morrell JM, Wallgren M. Removal of bacteria from boar ejaculates by Single Layer Centrifugation can reduce the use of antibiotics in semen extenders. Animal Reproduction Science 123, 64-69.

Van Wienen M, Johannisson, A., Wallgren, M, Joyce Parlevliet, J., Morrell JM. Single Layer Centrifugation with Androcoll™-P can be scaled-up to process larger volumes of boar semen. ISRN Veterinary Science doi.10.5402/2011/548385.

Ortega Ferrusola C, Johannisson A, Peña FJ, Tapia JA, Rodriguez-Martinez H, Dalin AM and JM Morrell (2011) Effect of different extenders and seminal plasma on the susceptibility of equine spermatozoa to lipid peroxidation after Single layer centrifugation (SLC), through Androcoll™-E. JEVS in press.

Morrell JM, Johannisson A, Rodriguez-Martinez H (2011) Effect of osmolarity and density of colloid formulations on the outcome of SLC-selection of stallion spermatozoa. ISRN Veterinary Science in press.

Colleoni, S., Lagutina, I., Rodriguez-Martinez, H., Lazzari, G., Galli C., Morrell, JM. New techniques for selecting spermatozoa for equine assisted reproduction. JEVS *in press*

Kruse, R, Dutta PC, Morrell JM. Colloid centrifugation removes seminal plasma and cholesterol from boar spermatozoa. Reproduction, Fertility and Development, *In press*

Blomqvist G, Persson M, Wallgren M, Wallgren P, Morrell JM. (2011) Removal of Virus from Boar Semen Spiked with Porcine Circovirus Type 2. Anim Reprod Sci *in press*

Morrell JM, Macias Garcia B, Pena FJ, Johannisson A. (2011) Processing stored stallion semen doses by Single Layer Centrifugation. Theriogenology, *in press.*

Hoogewijs M, Morrell JM, Van Soom A, Govaere J, Johannisson A, Piepers P, De Schauwer C, de Kruif A, De Vliegher S. (2011) Sperm selection using single layer centrifugation prior to cryopreservation can increase post thaw sperm quality in stallions. Equine Vet Journal, *In Press.*

In: Chemistry Researcher Biographical Sketches
Editors: E. S. Huseynov and V. Babayev
ISBN: 978-1-62100-771-5

Chapter 58

BIOGRAPHICAL SKETCH

Name:	Title:	Date of Birth:
Fathi Moussa	Paris Sud 11 University	1952/03/03

EDUCATION:

Institution and Location	Degree	Year Conferred	Scientific Field
Pharmacien Biologiste, Docteuren Sciences pharmaceutiques			

CONTACT POINTS:

IUT d'Orsay, Plateua de Moulon, 91400 Orsay, France

RESEARCH AND PROFESSIONAL EXPERIENCE:

Clinical Chemistry and Biomedical applications of fullerenes and carbon nanotubes.

Professional Appointments:

Fathi Moussa is Professor of Analytical Chemistry at Paris Sud 11 University and Clinical Biologist at Trousseau Hospital (Assistance Publique Hôpitaux de Paris)

Honors: Prix de l'Internat des Hôpitaux de Paris (1985), Prix de la Valorisation de la Recherche de l'Université Paris Sud (2004), Prix de la Chancellerie (2007).

Publications during last three years

1. N Gharbi, M Pressac, M Hadchouel, H Szwarc, S. R. Wilson and Fathi Moussa “[60] Fullerene is an in vivo Powerful Antioxidant With no Acute or Sub-acute Toxicity.” *Nano Letters* 2005, 5 (12): 2578 - 2585. (IF: 9.991)
2. J-Y Hautem, C Morel, R Couderc and F Moussa. “Liquid-chromatographic measurement of riboflavin and its coenzyme forms In human plasma and whole blood.” *Clin Chem.* 2006, 52: 907-908.
3. Séverine Ledoux, Simon Msika, Fathi Moussa, Etienne Larger , Philippe Boudou , Laurence Salomon , Carine Roy , Christine Clerici. "Comparison of nutritional consequences of conventional therapy of obesity, adjustable gastric banding, and gastric bypass". *Obesity Surgery* 2006, 16 (4): 414-414
4. Mhamed Ben Messaouda, Fathi Moussa, Bahoueddine Tangour, Henri Szwarc, Manef Abderrabba. “Addition Of Bio-Organic Compounds On C60: A Semi-Empirical Investigation of Its Reactivity With Glycine” Journal of Molecular Structure: THEOCHEM 2007, 809: 153–159
5. J. Desrivot, F. Moussa, P. Champy, A. Fournet, B. Figadère, C. Herrenknecht. “Development of a SPE/HPLC/DAD method for the determination of antileishmanial 2-substituted quinolines and metabolites in rat plasma”.. *Chromatogr. B* 2007, 854: 230–238.
6. Smaranda Leu ; Isabelle Arnulf ; Caroline Decaix ; Fathi Moussa ; Fabienne Clot ; Camille Boniol ; Yvan Touitou ; Richard Levy ; Marie Vidailhet ; Emmanuel Roze. Sleep and rhythm consequences of genetically induced loss of serotonin. SLEEP 2010, 33: 335-342.
7. Kolosnjaj-Tabi, J., Hartman, K.B., Boudjemaa, Sabah, Ananta, J.S., Morgant, G., Szwarc, H., Wilson, L.J. and Moussa, F. In Vivo Behavior of Large Doses of Ultrashort and Full-Length Single-Walled Carbon Nanotubes after Oral and Intraperitoneal Administration to Swiss Mice. ACS Nano 2010, 4 (3), pp 1481–1492.
8. Khair Elhareth; Christine Vauthier And Fathi Moussa. "Doxorubicin loading and in vitro release from poly(alkylcyanoacrylate) nanoparticles produced by Redox Radical Emulsion Polymerization". Journal of Applied Polymer Science 2011, 119, 2, 816-822.
9. Lagadu, Stéphanie; Lechevrel, Mathilde; Sichel, François; Breton, Jean; Pottier, Didier; Couderc, Rémy; Moussa, Fathi; Prevost, Virginie. "Increased levels of 8-oxodG, a biomarker of oxidative stress, in oesophageal cancer patients and the lack of its association with antioxidant vitamins and polymorphism in the hOGG1 and GST genes". Biomarkers (In press)
10. Szwarc, Henri; and Moussa, Fathi. Toxicity of [60] Fullerene: Confusion in the Scientific Literature. Jouranl of Nanoscience Letters (In press).

In: Chemistry Researcher Biographical Sketches
Editors: E. S. Huseynov and V. Babayev
ISBN: 978-1-62100-771-5

Chapter 59

BIOGRAPHICAL SKETCH

Name:	Title:	Date of Birth:
N. S. Muravyeva	PhD Geochemistry	06.01.1947

RESEARCH AND PROFESSIONAL EXPERIENCE:

37 years

Professional Appointments:

Senior researcher

Publications during last three years:

1. Муравьева Н.С., Сенин В.Г. Камафугиты Восточно-Африканского рифта: окислительно-восстановительные условия кристаллизации. Тезисы ЕСМПГ-2008, стр.53-54.

2. Муравьева Н.С., Сенин В.Г. Включения первичных карбонатов и барита в мантийных магмах Буньярунгуру (Восточно-Африканский рифт). "Геохимия магматических пород". Материалы XXV Всероссийского семинара с участием стран СНГ. Школа "Щелочной магматизм Земли"-Санкт-Петербург, май 2008 г.с.113-114.

3. Муравьева Н.С., Сенин В.Г. Камафугиты Восточно-Африканского рифта: оливин-шпинелевый мониторинг окислительно-восстановительных условий. "Геохимия магматических пород". Материалы XXV Всероссийского семинара с участием

стран СНГ. Школа "Щелочной магматизм Земли"-Санкт-Петербург, май 2008 г., с.114.

4. Муравьева Н.С., Сенин В.Г.Карбонат-силикатные равновесия в высокомагнезиальных ультракалиевых вулканитах Торо-Анколе (Восточно-Геохимия 2009, 9,с.937-957.

5. Muravyeva N.S., Belyatsky B.V. Petrology and geochemistry Toro Ankole kamafugite magmas: isotopic constraints. Geophysical Research Abstracts, Vol. 11, EGU2009-12651-1,2009EGUGeneralAssembly2009.

6. Muravyeva N.S., Belyatsky B.V. Petrology and geochemistry Toro Ankole kamafugite magmas: isotopic constraints. Geophysical Research Abstracts, Vol. 11, EGU2009-12651-1, 2009 EGU General Assembly 2009

7. Muravyeva N.S., Belyatsky B.V., Ivanov A.V. Geochemistry and petrology Toro Ankole kamafugite magmas: isotopic constraints. XXVI international conference Geochemistry of magmatic rocks, school "Geochemistry of alkaline rocks". Abstract volume. 2009, p.107-108.

8. Муравьева Н.С., Сенин В.Г. Камафугиты Восточно-Африканского рифта: окислительно-восстановительные условия кристаллизации. Электронный научно-информационный журнал "Вестник Отделения наук о Земле РАН" 2008, 1 URL: http://www.scgis.ru/russian/cp1251/h_dgggms/1-2008/informbul-1_2008/magm-23.pdf

9. Muravyeva N.S. Senin V.G. Kamafugites of the Eastern - African rift zone: redox states of crystallization. Электронный научно-информационный журнал "Вестник Отделения наук о Земле РАН" 2008, 1 URL: http://www.scgis.ru/russian/ cp1251/h_dgggms/1-2008/informbul-1_2008/magm-23e.pdf.

10. Лебедев Е.Б., Дернов-Пегарев В.Ф., Муравьева Н.С., Скрипник А.Я. Влияние физико - химических условий на аккумуляцию и сегрегацию металлических фаз в центробежном поле. Десятая международная конференция "Физико-химические и петрофизические исследования в науках о Земле" Москва 2009 Материалы конференции, с.225-228.

11. Muravyeva N.S., Belyatsky B.V. Petrology and geochemistry Toro Ankole kamafugite magmas: isotopic constraints. Geophysical Research Abstracts, Vol. 11, EGU2009-12651-1, 2009 EGU General Assembly 2009.

12. Muravyeva N.S., Belyatsky B.V., Ivanov A.V. Geochemistry and petrology Toro Ankole kamafugite magmas: isotopic constraints. XXVI international conference Geochemistry of magmatic rocks, school "Geochemistry of alkaline rocks". Abstract volume. 2009, p.107-108.

13. Муравьева Н.С., Беляцкий Б.В., ИвановА.В. Изотопно-геохимическая характеристика камафугитов Восточно-Африканского рифта. "Изотопные системы и время геологических процессов". Материалы IV Российской конференции по изотопной геохронологии. 2009 Санкт-Петербург Том II, с.34-37.

14. N. Muravyeva and V. Senin. Carbonate-Silicate Association in the Kamafugite of the Toro-Ankole Province (East African Rift) Geophysical Research Abstracts Vol. 12, EGU2010-11421-1, 2010 EGU General Assembly. 2010.

15. Муравьева Н. С., Беляцкий Б. В., Иванов А. В., Сенин В. Г. Геохимия и петрология ультракалиевых вулканитов Восточно-Африканской рифтовой зоны. "Магматизм и метаморфизм в истории Земли" Материалы XI Всероссийского петрографического совещания с участием зарубежных ученых. Екатеринбург, 2010. Том II, стр.81-82.

In: Chemistry Researcher Biographical Sketches
Editors: E. S. Huseynov and V. Babayev

ISBN: 978-1-62100-771-5

Chapter 60

BIOGRAPHICAL SKETCH

Name:	Title:	Date of Birth:
Alexandr P. Novikov	Ph.D	December 18, 1955

EDUCATION:

Institution and Location	Degree	Year Conferred	Scientific Field
Mendeleev Chemico-Technological University	M.Sc	1979.	
	Ph.D	1983	
	Full Doctor degree.	2004	

CONTACT POINTS:

119991, Kosygin str., 19, Moscow, Russian Federation.

RESEARCH AND PROFESSIONAL EXPERIENCE:

Membrane and chromatographic methods of actinide's partitioning. The field of his special interest is the behaviour of actinides in the environment. Dr. A.P. Novikov takes part in the State programs of remediation of Chelyabinsk and Krasnoyarsk areas of Russia. A.P. Novikov is a Member of Russian Scientific Council on Radiochemistry

AWARDS AND HONORS:

The premium of the Government of the Russian Federation on ecology (2007).

Professional Appointments:

1983-1988 - Vernadsky Insitute of Geochemistry and Analytical Chemistry, and occupied positions of Research Chemist () and 1988-2008 - Senior scientist
2008-to present - Head of Radiochemical laboratory

Publications during last three years:

Novikov A.P. Migration of Radioniclides in the Environment. // Geochemistry International, 2010. V.48. N13. P. 1-125.

Novikov A.P., Kalmykov S.N., Goryachenkova T.A., Myasoedov B.F. Colloidal transport of radionuclides in soils. J. of Russian Chem. Soc. 2010, V.54. N3. P.111-119. (in Russian)

Romanchuk Yu.A., Kalmykov S.N., Novikov A.P., Zaharova E.V. Sorption of actinides on mineral particles. J. of Russian Chem. Soc. 2010, V.54. N3. P.120-129. (in Russian)

Kuzovkina E.V., Malofeeva G.I., Petruhin O.M., Novikov A.P., Spivakov B.Ya. Recovery of Np(V), Pu(IV) and Am(III) by solid-phase extraction with supported N-benzoylphenylhydroxylamine. Radiochemistry. 2009 V.51, N 1, P. 26-30.

Goryachenkova T.A., Kazinskaya I.E., Kuzovkina E.V., Novikov A.P. Association of radionuclides with colloids in soil solution. Radiochemistry. 2009 V.51, N 2, P.201-211.

Novikov A.P., Lavrinovich E.A., Mogilevsky A.N., Fabelinsky Yu.A. Luminescence analysis of contaminated underground waters for the neptunium content and forms of occurrence. Radiochemistry. 2009 V.51, N 5, P. 537-642.

Novikov A.P., Kalmykov S.N., Goryachenkova T.A., Kazinskaya I.E., Kuzovkina E.V., Association of radionuclides with colloidal matter of underground water taken from observation wells in the zone of impact of lake Karachai. Radiochemistry. 2009 V.51, N 6, P. 644-649.

Novikov A.P., Kalmykov S.N., Goryachenkova T.A., Kazinskaya I.E., Kuzovkina E.V., Vlasova I.E., Tkachev V.V. Speciation of radionuclides in colloidal matter of underground water taken from observation wells in the zone of impact of lake Karachai. Radiochemistry. 2009 V.51, N 6, P. 649-654.

Novikov A.P., Kalmykov S.N., Myasoedov B.F., Ivanov I.A. Speciation of radionuclides in underground water taken from observation wells in the zone of impact of lake Karachai. Radiation safety, 2009, N1, P.12-22. (in Russian)

Alexander P. Novikov, Stepan N Kalmykov, Elena V Kuzovkina; Boris F Myasoedov, Kenso Fujiwara, Ai Fujiwara Evolution of actinide partitioning with colloidal matter collected at PA "Mayak" site as studied by sequential extraction J Radioanal. Nucl. Chem., 2009, V. 280, #3, p. 629-634.

In: Chemistry Researcher Biographical Sketches
Editors: E. S. Huseynov and V. Babayev

ISBN: 978-1-62100-771-5

Chapter 61

BIOGRAPHICAL SKETCH

Name:	Title:	Date of Birth:
Hiroaki Onoda	Kyoto Prefectural University	July 6, 1974

EDUCATION:

Institution and Location	Degree	Year Conferred	Scientific Field
Doctor of Engineering, Kobe University			

CONTACT POINTS:

Shimogamo-nakaragi, Sakyo, Kyoto 606-8522 Japan

RESEARCH AND PROFESSIONAL EXPERIENCE:

Inorganic Materials Chemistry

AWARDS AND HONORS:

2008 - The Award for Young Investigator of Japanese Association of Inorganic Phosphorus Chemistry

Publications during last three years:

Many papers in the field of inorganic materials.

In: Chemistry Researcher Biographical Sketches
Editors: E. S. Huseynov and V. Babayev
ISBN: 978-1-62100-771-5

Chapter 62

BIOGRAPHICAL SKETCH

Name:	Title:	Date of Birth:
Guangnan Ou	Ph D	Feb 22nd 1965

EDUCATION:

Institution and Location	Degree	Year Conferred	Scientific Field
School of Bioengineering, Jimei University	Ph D		

CONTACT POINTS:

Room 120-506, Cendong Road, Xiamen, P R China

RESEARCH AND PROFESSIONAL EXPERIENCE:

Guangnan Ou, male, received his B.Sc. in chemistry from Zhongshan University, China, in 1985, and his Ph.D. in physical chemistry from Xiamen University in 2007. In 1991, He joined the faculty of Jimei University where he is now associate professor of chemistry. His research interests include nonaqueous enzymatic catalysis

Professional Appointments:

Associate Professor

Publications during last three years

1. G. N. Ou, J. Yang, B. Y. He and Y. Z. Yuan. Buffer-mediated activation of *Candida antarctica* lipase B dissolved in hydroxyl-functionalized ionic liquids. *J. Mol. Catal. B: Enzym.*, 2011, 68 (1): 66-70.
2. G. N. Ou, L. Xu, B.Y. He and Y. Z. Yuan. Enhanced stability of charged dendrimer-encapsulated Pd nanoparticles in ionic liquids. *Chem. Commun.* 2008, 4210.
3. N. Ou, M. X. Zhu, J. R. She and Y. Z. Yuan. Ionic liquid buffers: a new class of chemicals with potential for controlling pH in non-aqueous media. *Chem. Commun.* 2006, 4626.

In: Chemistry Researcher Biographical Sketches
Editors: E. S. Huseynov and V. Babayev

ISBN: 978-1-62100-771-5

Chapter 63

BIOGRAPHICAL SKETCH

Name:	Title:	Date of Birth:
Byung-Dae Park	PH.D	06-09-1960

EDUCATION:

Institution and Location	Degree	Year Conferred	Scientific Field
DEPARTMENT OF WOOD SCIENCE AND TECHNOLOGY,	PH.D		

CONTACT POINTS:

1370 SANKYUG-DONG, BUK-KU, DAEGU, KOREA

RESEARCH AND PROFESSIONAL EXPERIENCE:

WOOD ADHESIVES, WOOD PANELS, THERMOSETS, NANOCOMPOSITES
Professional Appointments: ASSOCIATE PROFESSOR

PUBLICATIONS LAST 3 YEARS: ABOUT 15 SCIENTIFIC PAPERS

In: Chemistry Researcher Biographical Sketches
Editors: E. S. Huseynov and V. Babayev
ISBN: 978-1-62100-771-5

Chapter 64

BIOGRAPHICAL SKETCH

Name:	Title:	Date of Birth:
Jin Hee Park	Centre for Mined Land Rehabilitation (CMLR), The University of Queensland	09 Nov 1978

EDUCATION:

Institution and Location	Degree	Year Conferred	Scientific Field
Centre for Environmental Risk Assessment and Remediation, University of South Australia, Australia	Ph.D.,	08/2006 – 06/2010	
Environmental Geochemistry major, School of Civil, Urban and Geosystem Engineering Seoul National University, Seoul, Korea	M.Sc	03/2001 – 02/2003	
Department of Applied Biology and Chemistry Seoul National University, Seoul, Korea	B.Sc.,	03/1997 – 08/2000	

CONTACT POINTS:

Level 5, CMLR Sir James Foots Building (47A) Cnr College and Staff house Road, The University of Queensland, St Lucia Campus 4072, Australia

RESEARCH AND PROFESSIONAL EXPERIENCE:

04/2011 – Present - Postdoctoral Research Fellow at University of Queensland
08/2010 – 02/2011 - Research assistant at University of South Australia
04/2003 – 08/2006- Researcher at the Department of Analysis Research, Skin Research Institute, Amorepacific Corporation R&D center, Yongin, Korea
06/2002 – 09/2002 - Visiting researcher (Brain Korea 21 Abroad Research Program) Austrian Agency for Food and Safety, Vienna, Austria

Professional Appointments:

Postdoctoral Research Fellow

Publications during last three years:

Journal Paper

Jin-Hee Park, Panneer Panneerselvam, Dane Lamb, Girish Choppala, Nanthi Bolan, 2011, Role of organic amendments on enhanced bioremediation of heavy metal(loid) contaminated soils, Journal of Hazardous Materials, 185, 549-574, doi:10.1016/j.jhazmat.2010.09.082.

Jin-Hee Park, Nanthi Bolan, Mallavarapu Megharaj, Ravi Naidu, 2011, Isolation of phosphate solubilizing bacteria and their potential for lead immobilization in soil, Journal of Hazardous Materials, 185, 829-836, doi:10.1016/j.jhazmat.2010.09.095.

Jin-Hee Park, Nanthi Bolan, Mallavarapu Megharaj, Ravi Naidu, 2011, Comparative value of phosphate sources on the immobilization of lead, and leaching of lead and phosphorus in lead contaminated soils, Science of the Total Environment, 409, 853-860, doi:10.1016/j.scitotenv.2010.11.003.

Jin-Hee Park, Nanthi Bolan, Mallavarapu Megharaj, Ravi Naidu, 2011, Concomitant rock phosphate dissolution and lead immobilization by phosphate solubilizing bacteria (*Enterobacter* sp.), Journal of Environmental Management, 92, 1115-1120, doi:10.1016/j.jenvman.2010.11.031 .

Jinhee Park, Nanthi Bolan, Mallavarapu Megharaj, Ravi Naidu, 2011, Isolation of phosphate-solubilizing bacteria and characterization of their effects on lead immobilization, Pedologist 53, 67-75.

Nanthi Bolan, Ravi Naidu, Girish Choppala, Jinhee Park, Maria Luz Mora Dedik Budianta and Periyasamy Panneerselvam, 2011, Solute interactions in soils in relation to the bioavailability and environmental remediation of heavy metals and metalloids , Pedologist 53, 1-18.

Nanthi S. Bolan, Jin Hee Park, Brett Robinson, Ravi Naidu, Keun Young Huh, 2011, Phytostabilization: A green approach to contaminant containment, Advances in Agronomy,112

Jinhee Park, Nanthi Bolan, Mallavarapu Megharaj, Ravi Naidu, 2011, Relative value of phosphate compounds in reducing the bioavailability and toxicity of lead in contaminated soils, Water, Air & Soil Pollution. (Accepted)

Jin Hee Park, Nanthi Bolan, Mallavarapu Megharaj, Ravi Naidu, Jae Woo Chung, 2011, Bacterial-assisted immobilization of lead in soils: implications to remediation, Pedologist. 54, 162-174.

Jin Hee Park, Nanthi Bolan, Jae Woo Chung, 2011, Environmental monitoring of the role of phosphate compounds in enhancing immobilization and reducing bioavailability of lead in contaminated soils, Journal of Environmental Monitoring. (Accepted)

Conference Proceedings

Jinhee Park, Nanthi Bolan, Megharaj Mallavarapu and Ravi Naidu, Enhancing the solubility of insoluble phosphorus compounds by phosphate solubilizing bacteria, 19th World Congress of Soil Science, Soil Solutions for a Changing World, Brisbane, Australia (2010).

Jinhee Park, Nanthi Bolan, Megharaj Mallavarapu and Ravi Naidu, Effect of phosphate-induced immobilization of lead on its mobility and bioavailability, 19th World Congress of Soil Science, Soil Solutions for a Changing World, Brisbane, Australia (2010).

Jinhee Park, Nanthi Bolan, Megharaj Mallavarapu and Ravi Naidu, Isolation and characterization of phosphate solubilizing bacteria (PSB) and their effect on lead immobilization, East and Southeast Asia Federation of Soil societies (ESAFS), Seoul, Korea (2009).

Jinhee. Park, Megharaj Mallavarapu, Ravi Naidu and Nanthi Bolan, Immobilization of lead as affected by the solubility of phosphate fertilizers, Cleanup conference 09, Adelaide, Australia (2009).

In: Chemistry Researcher Biographical Sketches
Editors: E. S. Huseynov and V. Babayev
ISBN: 978-1-62100-771-5

Chapter 65

BIOGRAPHICAL SKETCH

Name:	Title:	Date of Birth:
Sang-Eon Park	Prof.	April 08, 1953

EDUCATION:

Institution and Location	Degree	Year Conferred	Scientific Field
Inha University	Ph.D		

CONTACT POINTS:

Department of Chemistry, Inha University
253 Yonghyun-dong, Nam-gu, Incheon, Korea 402-751
Phone: +82 32 860 7675

RESEARCH AND PROFESSIONAL EXPERIENCE:

N/A

AWARDS AND HONORS: N/A

Professional Appointments:

N/A

Publications during last three years:

N/A

In: Chemistry Researcher Biographical Sketches
Editors: E. S. Huseynov and V. Babayev
ISBN: 978-1-62100-771-5

Chapter 66

BIOGRAPHICAL SKETCH

Name:	Title:	Date of Birth:
Bertil Rolf Ragnar Persson (Persson BRR)	Lund University, Lund, SE 22185 Sweden	1938 October 12

EDUCATION:

Institution and Location	Degree	Year Conferred	Scientific Field
University of Lund, Sweden (Radiation Physics) Chemical engineer 1958 Technical College,	Doctor of Philosphy	1960-1978	

CONTACT POINTS:

Barngatan 2, SE 22185, LUND
Telephone: +46-(0)46-17 31 10 (office)
Mobile: +46(0)708278087
FAX: +46-(0)46-13 42 49
E-mail: Bertil_R.Persson@med.lu.se ;

RESEARCH AND PROFESSIONAL EXPERIENCE:

Environmental Radiological Health: My scientific career began 1963 with investigations of the health impact of the radioactive fall-out from atmospheric nuclear weapons tests in the food-chain lichenreindeer-man. Since 1972 I have been deeply engaged in gamma-spectrometry field studies around nuclear power plants in Sweden, and have studied the impact of various meteorological parameters on the recorded radiation field spectra. In 1980, 1989-90, 1994 and 1996 I participated in the Swedish Polar Expeditions to Arctic (North Pole September 10, 1996) and Antarctica (1989-90) with programs in marine radioecology and radio meteorology.

Environmental Sensor techniques: My polar reseach program was focused on sensors for measuring the Ozon layer, the cosmic radiation and earth´s magnetic field. These experiences were my contribution in the foundation of Tellusecure AB where we also are developing sensors for gas monitoring. Recently that research has conduced to ultra sensitive and specific gas sensors to be used for detection of narcotics in various objects and subjects.

Nuclear Medicine and Radiopharmaceutical: Radiochemical studies of short-lived technetium-99m started already in 1967 with the development of methods for preparations and quality control of several new radiopharmaceuticals (colloids, albumin, plasmin etc).

Medical Hyperthermia Therapy: A special computer controlled equipment for in vivo control of local hyperthermia in patients. This equipment was applied for use in combination with radiation therapy for treatment of superficial tumours. Microwave induced thermotherapy have been successfully applied in the treatment of benign prostate hyperplasia by using an interstitial microwave antenna inserted in a catheter.

Medical Nuclear Magnetic Resonance: I started already 1982 with biomedical applications of Nuclear Magnetic Resonance "NMR" and have since then developed and built special equipment for performing *in vivo* NMR studies in both animal and man .

Biomedical effects and Health impact of Electromagnetic Fields: Biomedical applications of NMR also involve the potential health hazards of clinical NMR-examinations and I have written extensive monographs on this subject. I also studied exposure to electromagnetic fields corresponding to those used in GSM mobile communications and found increased permeability of the blood brain barrier in the rat (BBB) to endogenous albumin. The interaction of static and extremely low frequency magnetic fields with calcium ion transport has been studied in normal and transformed human lymphocytes and in rat thymic cells human lymphocytes as well as Ca-45 transport in membrane vesicles of plant cells. Recent studies together with Belyaev et al at Stockholm University indicate that GSM microwaves do not induce DNA breaks or change chromatin, but affect expression of genes in rat brain under specific conditions of exposure. At present I am studying the use of pulsed electric fields for tumour and gene therapy and bioimpedance spectroscopy for dosimetry and tissue characterization.

Electro Chemo-Therapy (ECT): Since 1991 I am also engaged in using pulsed electric fields for tumour therapy with the objective to study its destructive effect on tumours and ability to introduce toxic compounds into tumour cells. Together with LG Salford we were the first to apply this method in the living brain. This novel mode of tumour treatment has been employed both for subcutaneous and coetaneous malignancies and the treatment of brain tumours with promising results. In electro chemotherapy treatment, only the cells exposed to a sufficiently strong electric field will respond immediately. The tumour cells that are permeabilized become much more accessible to hydrophilic molecules since these are normally rejected by the membrane barrier. Bleomycin, a rather non-toxic agent when present extra-cellular, but is very toxic anticancer agent when present intracellular has proven to be the most potent drug in ECT and is by far the one most often used. But cisplatin, another anticancer agent has also been found to be effective. The effect of electro-permeabilization on the cellular uptake of administered therapeutic agents is studied by using radioactive labelled compounds (111In , 99Tcm) and a scintillation camera.

Electro Immuno-Therapy (EIT): We have shown that the host's immune system is activated after tumour treatment with pulsed electric fields. ECT applied to liver tumour in the rat stimulates a response of the host's immune system against the tumour . Electro-permeabilization *in vivo* has recently shown to be very efficient in gene and protein delivery to living tissues. This can have very interesting implications in tumour gene therapy, but also in therapies for other diseases. Electro-chemotherapy and *in vivo* gene transfection are attractive treatment modalities for several reasons, besides the proven anti-tumour efficacy. The equipment is not very costly, the technical aspect of the therapy is uncomplicated and the treatment is generally rapidly performed.

Pulsed Electric Fields "PEF" combined with Radiation Therapy: Radiation therapy in combination with electro-permeabilization on subcutaneous rat glioma tumours has newly discovered to be an efficient therapy for highly resistant tumours. A distinct anti-tumour effect of the combined treatment of electric pulses and radiation treatment was observed. The discovery was made when sub-optimal radiation treatment was administered separately or in combination with electric pulses of high voltage to subcutaneous rat brain tumours. The treatment was repeated four consecutive days and evaluated by tumour growth delay TGD and microscopic examination. An effect on the tumour vasculature was investigated by staining the excised tumours for Factor VIII/von Willebrand Factor.

Pulsed Electric Fields "PEF" combined with Thermotherapy: The electric pulses used in electro-permeabilization *in vivo* have recently been shown to affect the vasculature of tumours and tissues in mainly an obstructive way. Inhibition of the blood perfusion in tumour and normal tissue can thus be valuable in applications where controlling the blood flow is desired. This is of great importance in hyperthermia and thermotherapy in well perfused tissue and tumours. There is also reason to believe that treatment with pulsed electric fields enhance the release of heat shock proteins that might cause an enhanced immune reaction.

Radio and immunotherapy of brain tumours: The most malignant brain tumours, the malignant glioma (also named glioblastoma multiforme –GBMor astrocytoma grade IV) are among the most therapy-resistant human cancers. The fact that no treatment hitherto has been successful may be explained by the way the malignant glioma grows in the brain: outwards from a centre, which by time turns into necrosis. The proliferating tumour front grows by progressively killing the surrounding normal brain cells, neuronal and glial. The tumour has an irregular shape, often with arms protruding into the brain in an octopus-like growth. Further, the tumour sends migrating tumour cells into the surrounding brain. We name these cells "guerrilla cells" as they hide in the normal cell population around the tumour. The key for success in the treatment of glioblastoma is a method to reach the migrating "guerrilla-cells". Radiotherapy and chemotherapy have hitherto proven not to be efficient. The aim of this project is to develop new radiation and immunotherapy modalities for brain tumours.

AWARDS AND HONORS:

The 2010 Most Influential Bioelectromagnetics Journal Paper by Citation Award from among the primary research articles published between 2005 and 2009.

Professional Appointments:

July 1980 - Oct 2005: Full Professor in Medical Radiation Physics at University of Lund, Lund University, Faculty of Medicine, PO Box 1703, S-22101 LUND, Sweden together with position as fully qualified hospital physicist at the Department of Radiation Physics, University Hospital, S-221 85 LUND.
January 1980 - July 1980: Full Professor in Radioecology at Swedish Agricultural University, S-750 07 Uppsala, Sweden
January 1975 - July 1980: Special Research professor in radioecology at Swedish Natural Research CouncilBox 6711, S-11385 Stockholm, Sweden
July 1970 - January 1975: Associate professor in radiation Physics at University of Lund, Department of radiation Physics, Lasarettet, S-221 85 LUND, Sweden
June 1969 - July 1970: Hospital physicist at Lund University Hospital, Department of Hospital Physics, Lasarettet, S-221 58 LUND,

Publications during last three years:

2011

1. Persson B R R and Holm E. (2011) Polonium-210 and Lead-210 in the Terrestrial Environment. Journal of Environmental Radioactivity 2102:420-429.

2010

2. B. R. R. Persson, C. B. Koch, G. Grafstrom, C. Ceberg, P. M. Af Rosenschold, H. Nittby, B. Widegren and L. G. Salford. "Radiation Immunomodulatory Gene Tumor Therapy of Rats with Intracerebral Glioma Tumors". Radiation Research 173(4) (2010) 433-440
3. Persson B R R, Bauréus Koch C, Grafström G, Ceberg C, Nittby H, Widegren B and Salford L G. (2010) Tumour response from immunization with interferon-gamma (IFNg) transfected syngeneic tumour cells combined with radiation therapy in rats with contra-lateral tumours. Cancer Therapy 8A (on line www.cancer-therapy.org).
4. Grafström G, Göransson I, Kaviani Moghadam M, Persson B R R, Magnusson M, Höglund P, Johansson K, Jönsson C, Killander F, Malmström P, Weddig A and Kjellén E. (2010). Electrical Impedance Spectrometry (EIS) evaluation of the outcome of Pentoxifylline and vitamin E treatment for prevention of radiation-induced side-effects in women with breast cancer: . In Annals of Medical Radiation Physics. 22185 Lund, Sweden.

2009

5. Belyaev I Y, Markova E, Hillert L, Malmgren L O G and Persson B R R. (2009) Microwaves From UMTS/GSM Mobile Phones Induce Long-Lasting Inhibition of 53BP1/gamma-H2AX DNA Repair Foci in Human Lymphocytes. Bioelectromagnetics 30(2),129-141.
6. Nittby H, Salford L G, Grafstrom G, Brun A, Malmgren L, Persson B R R and Eberhardt J. (2009) Response to Comments on Cognitive Impairment in Rats After Long-Term Exposure to GSM-900 Mobile Phone Radiation by Nittby et al. (Bioelectromagnetics 29:219-232, 2008). Bioelectromagnetics 30(6),509-509.

2008

7. Eberhardt J L, Persson B R R, Brun a E, Salford L G and Malmgren L O G. (2008) Blood-brain barrier permeability and nerve cell damage in rat brain 14 and 28 days after exposure to microwaves from GSM mobile phones. Electromagnetic Biology and Medicine 27(3),215-229.
8. Grafstrom G, Nittby H, Brun A, Malmgren L, Persson B R R, Salford L G and Eberhardt J. (2008) Histopathological examinations of rat brains after long-term exposure to GSM-900 mobile phone radiation. Brain Research Bulletin 77(5),257-263.
9. Nittby H, Grafstrom G, Eberhardt J L, Malmgren L, Brun A, Persson B R R and Salford L G. (2008) Radiofrequency and extremely low-frequency electromagnetic field effects on the blood-brain barrier. Electromagnetic Biology and Medicine 27(2),103-126.
10. Nittby H, Grafstrom G, Tian D P, Malmgren L, Brun A, Persson B R R, Salford L G and Eberhardt J. (2008) Cognitive impairment in rats after long-term exposure to GSM-900 mobile phone radiation. Bioelectromagnetics 29(3),219-232.
11. Persson B R R, Koch C B, Grafstrom G, Ceberg C, Af Rosenschold P M, Widegren B and Salford L G. (2008). Survival of rats with N29 brain tumours after irradiation with 5 or 15 Gy and immunization with IFN-gamma secreting tumour cells. In 1st International Conference on Biomedical Engineering and Informatics, ed. Peng Y Z Y. Sanya, PEOPLES R CHINA: Ieee Computer Soc.

2007

12. Salford L G, Nittby H, Brun A, Grafstrom G, Malmgren L, Sommarin M, Eberhardt J, Widegren B and Persson B R R. (2007). The mammalian brain in the electromagnetic fields designed by man with special reference to blood-brain barrier function, neuronal damage and possible physical mechanisms. In 22nd Nishinomiya-Yukawa Memorial Symposium. Kyoto, JAPAN: Progress Theoretical Physics Publication Office.
13. Salford L G, Nittby H, Brun A, Grafström G, Eberhardt J, Malmgren L and Persson B R R. (2007) Nonthermal effects of EMF upon the mammalian brain. Environmentalist DOI 10.1007/s10669-007-9118-4,7.
14. Salford L G, Nittby H, Krogh M, Grafström G, Berlin H, Rehn G, Eberhardt J, Malmgren L, Persson B R R and Widegren B. (2007) Exposure to Global System for Mobile Communications at 1800 MHz Significantly Changes Gene Expression in Rat Hippocampi and Cortex.

In: Chemistry Researcher Biographical Sketches
Editors: E. S. Huseynov and V. Babayev ISBN: 978-1-62100-771-5

Chapter 67

BIOGRAPHICAL SKETCH

Name:	Title:	Date of Birth:
Eva Petrovova	Doctor of veterinary medicine	19.11.1979

EDUCATION:

Institution and Location	Degree	Year Conferred	Scientific Field
Department of Biology and Genetics, University of Veterinary Medicine and Pharmacy, Kosice, Slovak Republic			

CONTACT POINTS:

University of Veterinary Medicine and Pharmacy, Komenskeho 73, 041 81 Kosice, Slovak Republic

RESEARCH AND PROFESSIONAL EXPERIENCE:

Environmental and Developmental Toxicology
Professional Appointments: university lecturer

AWARDS AND HONORS:

2nd place in the competition of young researchers of Slovak agency of agricultural sciences

Publications during last three years:

Petrovova, E., Mazensky, D., Vdoviakova, K., Massanyi, P., Luptakova, L., Smrco, P., 2010: Effect of bendiocarb development of the chick embryo. J. Appl. Toxicol., 30, 397-401.

Petrovova, E., Mazensky, D., Luptakova, L., Holovska, K., Spalekova, E., Massanyi, P., Haladova, E., Tóth, T., 2010: Alterations in the rabbit lymphoid tissue after bendiocarb administration. Journal of Environmental Science and Health Part B, 45, 719-728.

Capcarova, M., Petrovova, E., Flesarova, S., Dankova, M., Massanyi, P., Danko, J., 2010: Bendiocarbamate induced alterations in selected parameters of rabbit homeostasis after experimental peroral administration. Pesticide Biochemistry and Physiology, 98, 213-218.

Luptakova, L., Balent, P., Valencakova, A., Hisira, V., Petrovova, E., 2010: Detection of Toxoplasma gondii and Encephalitozoon spp. In wild boars by serological and molecular methods. Revue Med. Vet., 161, 559-563.

Mazensky, D., Radonak, J., Danko, J., Petrovova, E., Frankovicova, M., 2010: Anatomical study to blood supply to the spinal cord in the rabbit. Spinal Cord, online, in press

Petrovova, E., Sedmera, D., Lesnik, F., Luptakova, L., 2009: Bendiocarb effect on liver and central nervous system in the chick embryo. J. Environ. Sci. Health, Part B, 44, 4, 383-388.

Valencakova, A., Balent, P., Petrovova, E., Novotny, F., Luptakova, L., 2008: Encephalitozoonosis in household pet Nederland Dwarf rabbits (Oryctolagus cuniculus). Veterinary Parasitology, 153, 265-269.

In: Chemistry Researcher Biographical Sketches
Editors: E. S. Huseynov and V. Babayev
ISBN: 978-1-62100-771-5

Chapter 68

BIOGRAPHICAL SKETCH

Name:	Title:	Date of Birth:
Matthew Picklo	Supervisory Research Physiologist- Research Leader	

EDUCATION:

Institution and Location	Degree	Year Conferred	Scientific Field
University of Delaware, Newark, DE	B.A.	1990	Biological Sciences
Vanderbilt University	Ph.D.	1995	Pharmacology

CONTACT POINTS:

N/A

RESEARCH AND PROFESSIONAL EXPERIENCE:

AWARDS AND HONORS:

1990-1993 - NIH Pharmacology Training Grant
1993-1995 - NIH SCOR Hypertension Training Grant
1995-1996 - Special Trustees Award of Middlesex Hospital, London UK Redirecting the Tropism of HSV Vectors"
1998 - Young Investigator Award Albany Mitochondrial Conference
1998 - Young Investigator Award The Oxygen Society Annual Meeting
1998-2001 - NRSA ES05826-01, "Toxicity of Catechol Thioether Adducts"
2008 - Hermann Esterbauer Award, HNE-Club, International Society for Free Radical Research

Professional Appointments:

1995-1996 - Post-doctoral fellow University College, London, UK
1996-2001 - Post-doctrol fellow Vanderbilt University Dept of Pathology, Nashville, TN
2001-2007 - Assistant Professor, University of North Dakota, Dept of Pharmacol, Physiol, and Therapeut.
2007-2009 - Associate Professor, University of North Dakota, Dept of Pharmacol, Physiol, and Therapeut.
2007-2008 - Associate Editor, *Journal of Alzheimer's Disease*
2009-present - Research Leader, USDA-ARS Grand Forks Human Nutrition Research Center
2006-present - Associate Editor, *Lipids*

Publications during last three years:

Peer-Reviewed Research

1. Picklo MJ, Wiley RG, Lappi D, Robertson D. Noradrenergic lesioning with an anti-dopamine beta-hydroxylase immunotoxin. *Brain Research* 1994 (666) 195-200.
2. Picklo MJ, Wiley RG, Lonce S, Lappi DA, Robertson D. Anti-dopamine beta-hydroxylase immunotoxin-induced sympathectomy in adult rats. *The Journal of Pharmacology and Experimental Therapeutics* 1995 (275) 1003-1010.
3. Wrenn CC, Picklo MJ, Lappi DA, Robertson D, Wiley RG. Central noradrenergic lesioning using anti-DBH-saporin: anatomical findings. *Brain Research* 1996 (740) 175-184.
4. Montine TJ, Picklo MJ, Amarnath V, Amarnath K, Whetsell WO, Graham DG. Neurotoxicity of endogenous cysteinylcatechols. *Experimental Neurology* 1997 (148) 26-33.
5. Picklo MJ, Amarnath V, McIntyre, JO, Graham DG, Montine TJ. 4-Hydroxy-2(*E*)-nonenal inhibits CNS mitochondrial respiration at multiple sites. *The Journal of Neurochemistry* 1999 (72) 1617-1624.
6. Picklo MJ, Amarnath V, Graham DG, Montine TJ. Catechol thioethers may be pro-oxidant or antioxidant. *Free Radical Biology and Medicine* 1999 (27) 271-277.
7. Picklo MJ, Zhang J, Nguyen V, Graham DG, Montine TJ. HPLC quantitation of cytochrome c using 393 nm detection. *Analytical Biochemistry* 1999 (276) 166-170.
8. Zhang J, Kravistov A, Amarnath V, Picklo MJ, Graham DG, Montine TJ. Catechol mercapturates increase dopamine-induced apoptosis: potential relevance to Parkinson's disease. *The Journal of Neurochemistry* 2000 (74) 970-978.
9. Neely MD, Zimmermann L, Picklo MJ, Ou, JJ, Morales CR, Montine KS, Amarnath V, Montine TJ. Congeners of N-acetyl-L-cysteine but not aminoguanidine act as neuroprotectants from the lipid peroxidation product 4-hydroxy-2-nonenal. *Free Radical Biology and Medicine* 2000 (29) 1028-1036.
10. Picklo MJ and Montine TJ. Acrolein inhibits respiration in isolated brain mitochondria. *Biochimica et Biophysica Acta 2001* (1535) 145-152.

11. Picklo MJ, Olson S, Markesbery W, Montine TJ. Expression and activities of aldo-keto oxidoreductases in Alzheimer's disease. *Journal of Neuropathology and Experimental Neurology* 2001 (60) 686-695.
12. Picklo MJ, Olson S, Hayes JD, Markesbery W, Montine TJ. Elevation of AKR7A2 (succinic semialdehyde reductase) in neurodegenerative disease. *Brain Research* 2001 (916) 229-238.
13. Sidell KR, Montine KS, Picklo MJ, Olson SJ, Amarnath V, Montine MJ. Mercapturate metabolism of 4-hydroxy-2-nonenal in rat and human cerebrum. *Journal of Neuropathology and Experimental Neurology*. 2003 (62) 146-153.
14. Nguyen E, and Picklo MJ. Inhibition of succinic semialdehyde dehydrogenase by alkenal products of lipid peroxidation. *Biochimica et Biophysica Acta : Molecular Basis of Disease*. 2003 (1637) 107-112.
15. Murphy T, Amarnath V, Picklo MJ. Oxidation of 4-hydroxynonenal in rat brain slices. *Chemico-Biological Interactions*. 2003 (143-144) 101-105.
16. Murphy T, Amarnath V, Picklo MJ. Mitochondrial oxidation of 4-hydroxy-2-nonenal in rat cerebral cortex. *Journal of Neurochemistry*. 2003 (84) 1313-21.
17. Murphy T, Amarnath V, Gibson KM, and Picklo MJ. Oxidation of 4-hydroxy-2-nonenal by succinic semialdehyde dehydrogenase (ALDH5A). *Journal of Neurochemistry*. 2003 (86) 298-305.
18. Kubatova A, Steckler TS, Gallagher JR, Hawthorne SB, and Picklo MJ. Toxicity of wide-range polarity fractions from wood smoke and diesel exhaust particulate obtained using subcritical water. *Environmental Toxicology and Chemistry*. 2004 (23) 2243-2250.
19. Aberle NS, Picklo MJ, Amarnath V, Ren J. Inhibition of cardiac myocyte contraction by 4-hydroxy-*trans*-2-nonenal. *Cardiovascular Toxicology*. 2004 (4) 21-28.
20. Murphy T, Poppe C, Porter J, Montine TJ, and Picklo MJ. 4-Hydroxy-*trans*-2-nonenoic acid is a GHB receptor ligand in the cerebral cortex and hippocampus. *Journal of Neurochemistry*. 2004 (89) 1462-70.
21. Meyer MJ, Mosely DM, and Picklo MJ. Metabolism of 4-hydroxy-*trans*-2-nonenal by CNS mitochondria is dependent on age and NAD^+ availability. *Chemical Research in Toxicology*. 2004(17) 1272-79.
22. Murphy T, Arntzen R, Picklo MJ. Nitrate-based vasodilators inhibit multiple vascular aldehyde dehydrogenases. *Cardiovascular Toxicology*. 2005 (5) 321-332.
23. Honzatko A, Brichac J, Murphy TC, Reberg A, Kubatova A, Smoliakova IP, Picklo MJ. Enantioselective metabolism of *trans*-4-hydroxy-2-nonenal by brain mitochondria. *Free Radical Biology and Medicine*. 2005 (39), 913-24.
24. Struys EA, Verhoeven NM , Jansen EEW , ten Brink HJ, Gupta M, Burlingame TG , Quang LS, Maher T, Rinaldo P, Snead OC, Goodwin AK, Weerts EM, Murphy TC, Picklo MJ, Jakobs C, and Gibson KM. Metabolism of gamma-hydroxybutyrate to D-2-hydroxyglutarate in mammals: further evidence for D-2-hydroxyglutarate dehydrogenase. *Metabolism: Clinical and Experimental*. 2006 (53), 353-358.
25. Kubatova A, Dronen L, Picklo MJ, Hawthorne SB. Midpolarity and nonpolar wood smoke PM fractions deplete glutathione in RAW 264.7 macrophages. *Chemical Research in Toxicology*. 2006 (19) 255-261.

26. Kubatova A, Murphy T, Combs C, Picklo MJ. Astrocytic biotransformation of trans-4-hydroxy-2-nonenal is dose-dependent. *Chemical Research in Toxicology.* 2006 (19) 844-51.
27. Milne G, Morrow JD, Picklo MJ. Elevated oxidation of docosahexaenoic acid, 22:6 (n-3), in brain regions of rats undergoing ethanol withdrawal. *Neuroscience Letters.* 2006 (405) 172-174.
28. Leiphon LJ and Picklo MJ. Inhibition of aldehyde detoxification in CNS mitochondria by fungicides. *Neurotoxicology.* 2007 (28) 143-149.
29. Grimsrud P, Picklo MJ, Griffin TJ, Bernlohr D. Carbonylation of adipose proteins in obesity and insulin resistance: Identification of adipocyte fatty acid-binding protein as a cellular target of 4-hydroxynonenal. *Molecular and Cellular Proteomics.* 2007 (6) 624-637.
30. Brichac J, Honzatko A, Picklo MJ. Direct and indirect high-performance liquid chromatography enantioseparation of *trans*-4-hydroxy-2-nonenoic acid. *Journal of Chromatography A* (2007) 1149(2) 305-311.
31. Brichac J, Ho KK, Honzatko A, Wang R, Lu X, Weiner H, and Picklo MJ. Enantioselective Oxidation of *Trans*-4-Hydroxy-2-Nonenal is ALDH Isozyme and Mg^{2+}-Dependent. *Chemical Research in Toxicology.* 2007 (20) 887-895.
32. Honzatko A, Brichac J, and Picklo MJ. LC-ESI-MS/MS analysis of *trans*-4-hydroxy-2-nonenal enantiomers and metabolites. *Journal of Chromatography B.* 2007 857(1):115-22.
33. Whittsett J, Picklo MJ, Vasquez-Vivar J. 4-hydroxy-2-nonenal increases superoxide anion radical in endothelial cells via stimulated GTPcyclohydrolase-1 proteasomal degradation. *Atherosclerosis, Thrombosis, and Vascular Biology.* 2007 27(11):2340-7.
34. Long EK, Murphy TC, Leiphon LJ, Watt J, Morrow JD, Milne GL, Howard JRH, Picklo MJ. *Trans*-4-Hydroxy-2-hexenal is a neurotoxic product of docosahexaenoic (22:6; *n*-3) acid oxidation. *Journal of Neurochemistry.* 2008 105(3):714-24.
35. Long EK, Smoliakova I, Honzatko A, and Picklo MJ. Structural Identification of α,β-Unsaturated Aldehydes by GC/MS is Dependent upon Ionization Method. *Lipids.* 2008 43(8):765-74.
36. Swenby NP and Picklo MJ. The conserved R166 residue of ALDH5A (succinic semialdehyde dehydrogenase) has multiple functional roles. *Chemico-Biological Interactions.* 2009 178(1-3):70-4
37. Picklo MJ. Ethanol intoxication increases hepatic *N*-lysyl protein acetylation. *Biochemical Biophysical Research Communications.* 2008 Nov 21;376(3):615-9.
38. Long EK, Rosenberger TA, Picklo MJ. Ethanol withdrawal increases glutathione adducts of 4-hydroxy-2-hexenal but not 4-hydroxyl-2-nonenal in the rat cerebral cortex. *Free Radical Biology and Medicine.* 2010 48(3):384-90.
39. Picklo MJ, Azenkang A, Hoffmann M. Trans-4-oxo-2-nonenal potently alters mitochondrial function. Free Radical Biology and Medicine. 2011Jan 15;50(2):400-7.
40. Gonnella TP, Leedahl T, Picklo MJ. NADH fluorescent analysis of the effects of magnesium ions on ALDH2. *Chemico-Biological Interactions. In press*
41. Uthus EO and Picklo MJ. Methionine Sulfoxide Disposition is Altered in Animal Models of Obesity. Submitted.

Other Peer-Reviewed Publications

1. Picklo MJ. Methods of sympathetic degeneration and alteration. *The Journal of the Autonomic Nervous System* 1997 (62) 111-125.
2. Picklo MJ, Montine TJ, Amarnath, V, and Neely MD. Carbonyl toxicology and Alzheimer's disease. *Toxicology and Applied Pharmacology* 2002 (184) 187-197.
3. Kubatova A, Honzatko A, Brichac J, Long E, Picklo MJ. Analysis of HNE metabolism in CNS models. *Redox Report.* invited review. 2007 (12) 16-19.
4. Picklo MJ and Montine TJ. Mitochondrial Effects of Lipid-Derived Neurotoxins. Invited review. *Journal of Alzheimer's Disease.* 2007; 12(2) 185-93.
5. Malaspina P, Picklo MJ, Jakobs C, Snead OC, and Gibson KM. Comparative Genomics of Aldehyde Dehydrogenase 5A1 (succinate semialdehyde dehydrogenase) and accumulation of gamma-hydroxybutyrate associated with its deficiency. *Human Genomics.* 2009 Jan;3(2):106-20.
6. Long EK and Picklo MJ. "*Trans*-4-hydroxy-2-hexenal, a product of *n*-3 fatty acid peroxidation: Make some room HNE...". *Free Radical Biology and Medicine.* Invited Review. 2010 Jul 1;49(1):1-8.

Publications

1. Robertson D, Beck C, Gary T, Picklo M. Classification of autonomic disorders. *International Angiology* 1993 (12) 93-102.
2. Robertson D, Picklo MJ. Gene therapy enters the clinical arena. *Drug Therapy* 1993 (10), 71-74.
3. Picklo MJ. Methods of sympathetic degeneration and alteration. *The Journal of the Autonomic Nervous System* 1997 (62) 111-125.
4. Montine TJ, Amarnath V, Picklo MJ, Sidell KR, Zhang J, Graham DG. Dopamine mercapturate can augment dopaminergic neurodegeneration. *Drug Metabolism Reviews* 2000 (32) 363-376.
5. Graham DG, Picklo MJ, Zhang J, Montine TJ. Roles of Quinones in catechol toxicity. in *The Role of Catechol Quinone Species in Cellular Toxicity;* 1999 Editor C.R. Creveling;. F.P.Graham.
6. Montine TJ, Amarnath V, Picklo MJ, Sidell KR, Zhang J, Graham DG. Endogenous brain catechol thioethers in dopaminergic degeneration. in *Neurotoxic Factors in Parkinson's Disease and Related Disorders;* 2001 Eds. A.Storch and M.A. Collins. Kluwer Academic.
7. Picklo MJ. Autonomic Disorders in Animals. in *Primer of the Autonomic Nervous System;* 2003 Eds. D Robertson, P Low, G Burnstock, and I Biaggioni. Academic Press.
8. Montine KS, Sidell KR, Picklo MJ, and Montine TJ. Metabolism of endogenous neurotoxins in neurodegenerative diseases. *Recent Research Developments in Human Pathology.* 2003 (1) 63 - 76.
9. Poppe C, Murphy TC, Montine TJ, Picklo MJ. 4-Hydroxy-*trans*-2-nonenal is oxidized by ALDH3A and ALDH5A in the human brain. Proceedings of the 12th Meeting on the Enzymology and Molecular Biology of Carbonyl Metabolism. Purdue Press. 2005.

10. Picklo, MJ. The disposition of lipid-derived carbonyls in Alzheimer's disease. in Oxidative Stress and Age-Related Neurodegeneration; 2005 Eds. Y Luo and L Packer. Taylor and Francis Press.
11. Kubatova A, Honzatko A, Brichac J, Long E, Picklo MJ. Analysis of HNE metabolism in CNS models. *Redox Report.* invited review. 2007 (12) 16-19.
12. Brichac J, Honzatko A, Leiphon L, and Picklo MJ. *Trans*-4-hydroxy-2-nonenal (HNE), a biochemical tool (...as well as a toxin). Proceedings of the 13th Meeting on the Enzymology and Molecular Biology of Carbonyl Metabolism. Purdue Press. 2007.
13. Picklo MJ. "So you were volunteered to be a graduate director..." Young Chemists Committee Newsletter, American Chemical Society, 2009.
14. Picklo MJ. "Anti-oxidants in food: what are they, how are they measured, and what are they really doing?" Grand Forks Herald, Grand Forks, ND. 2010.

In: Chemistry Researcher Biographical Sketches
Editors: E. S. Huseynov and V. Babayev
ISBN: 978-1-62100-771-5

Chapter 69

BIOGRAPHICAL SKETCH

Name:	Title:	Date of Birth:
Carolina Henritta Pohl	PhD (Microbiology)	23 April 1972

EDUCATION:

Institution and Location	Degree	Year Conferred	Scientific Field
Department of Microbial, Biochemical and Food Biotechnology, University of the Free State			

CONTACT POINTS:

PO Box 339
University of the Free State
Bloemfontein, 9300
South Africa

RESEARCH AND PROFESSIONAL EXPERIENCE:

Bioactive lipids in fungi

Professional Appointments:

Researcher: Jan 2005-June 2006
Lecturer: June 2006-January 2008
Senior Lecturer: January 2008-present

Publications during last three years:

Ells R, Kock JLF, Van Wyk PWJ, Botes PJ & Pohl CH (2009) Arachidonic acid increases antifungal susceptibility of Candida albicans and Candida dubliniensis. Journal of Antimicrobial Chemotherapy 63: 124-128.<?xml:namespace prefix = o ns = "urn:schemas-microsoft-com:office:office" />

Kock JLF, Swart CW, Ncango DM, Kock JLF (Jr), Munnik IA, Maartens MMJ, Pohl CH & van Wyk PWJ (2009) Development of a yeast bio-assay to screen anti-mitochondrial drugs. Current Drug Discovery Technologies 6: 186-191.

Leeuw NJ, Swart, CW, Ncango DM, Kriel WM, Pohl CH, van Wyk PWJ, & Kock JLF (2009) Anti-inflammatory drugs selectively target sporangium development in *Mucor. Canadian Journal of Microbiology* 55: 1392-1396.

Ncango DM, Swart CW, Pohl CH, van Wyk PWJ & Kock JLF (2010) Mitochondrion activity and dispersal of *Aspergillus fumigatus and Rhizopus oryzae.* African Journal of Microbiology Research 4(9): 830-835.

Thibane VS, Kock JLF, Ells, R, Van Wyk PWJ & Pohl CH (2010) Effect of marine polyunsaturated fatty acids on biofilm formation of Candida albicans and Candida dubliniensis. Marine Drugs (Special Edition: Marine Lipids) 8: 2597-2604.

Swart CW, Van Wyk PWJ, Pohl CH & Kock JLF (2010) Variation in mitochondrial activity over the life cycle of *Nadsonia fulvescence.* African Journal of Microbiology Research 4: 1727-1732.

Leeuw NJ, Swart CW, Joseph M, Pohl CH, Van Wyk PWJ, Coetsee E, Swart HC, Hugo A & Kock JLF (2010) The effects of palm oil breakdown products on lipid turnover and morphology of fungi. Canadian Journal of Microbiology 56: 883-889.

Swart CW, Swart HC, Coetsee E, Pohl CH, Van Wyk PWJ & Kock JLF (2010) 3-D architecture and elemental composition of fluconazole treated yeast asci. Scientific Research and Essays 5: 4311-3417.

Ells R, Kock JLF, Albertyn J, Kemp G & Pohl CH (2011) Effect of inhibitors of arachidonic acid metabolism on prostaglandin E_2 production by *Candida albicans and Candida dubliniensis* biofilms. Medical Microbiology and Immunology 200: 23-28.

Ells R, Kock JLF & Pohl CH (2011) Review: *Candida albicans or Candida dubliniensis.* Mycoses 54: 1-16.

Pohl CH, Smit MS & Albertyn J. *Rhodotorula bloemfonteinensis* sp. nov., *Rhodotorula eucalyptica* sp. nov., *Rhodotorula orientis* sp. nov. and *Rhodotorula pini* sp. nov., yeasts isolated from monoterpene rich environments. International journal of Systematic and Evolutionary Microbiology (accepted for publication).

Swart CW, Pohl CH & Kock JLF. The anti-motochondrial antifungal assay for the discovery and development of new drugs. expert Opinion on Drug Discovery (accepted for publication).

In: Chemistry Researcher Biographical Sketches
Editors: E. S. Huseynov and V. Babayev

ISBN: 978-1-62100-771-5

Chapter 70

BIOGRAPHICAL SKETCH

Name:	Title:	Date of Birth:
Juan Carlos Poveda Jaramillo	Laboratorio de Colisiones Atómicas Moleculares, Instituto de Ciencias Físicas, Universidad Nacional Autónoma de México	May 6 of 1972

EDUCATION:

Institution and Location	Degree	Year Conferred	Scientific Field
B.Sc. in Chemistry at Universidad Industrial de Santander, Colombia; M.Sc. in Chemistry at Universidad Industrial de Santander, Colombia; Doctor in Chemical Sciences at Universidad Nacional Autónoma de México, México; Postdoctoral Fellow at CONACyT-Universidad Nacional Autónoma de México, México			

CONTACT POINTS: N/A

RESEARCH AND PROFESSIONAL EXPERIENCE:

Research and Professional Experience: Dr. Poveda have a long experience in spectroscopy techniques. The main goals are related with the spectroscopy of cooled molecules in molecular beams and related techniques to obtain information about the absorption and emission radiation, and ionization and dissociation processes that takes place at these experimental conditions. Also, the research work had been addressed to develop experimental setups to improve the quality of collected data, and include thermal desorption and laser desorption systems to the analysis of molecular clustering during the vaporization processes. Dr. Poveda combine data from experiments and theoretical quantum calculations at high level of theory to do a comprehensive analysis of molecular behavior under different experimental environments simulating real situations.

Professional Appointments:

Dr. Poveda had been focused to use the different spectroscopic techniques to the analysis of real problems in organic synthesis, analytical chemistry, petroleum chemistry and plasma physics. The main techniques which are used combination with theoretical data are 1D and 2D NMR spectroscopy, infrared spectroscopy, mass spectrometry using different ionization techniques, and analyzers as ToF, quadrupoles, electrostatic and magnetic. And laser techniques for optical spectroscopies, which can include mass spectrometry and photoelectron spectroscopy. In recent years the Dr. Poveda has specialized in the generation and manipulation of ion beams.

Publications during last three years:

Multiphoton ionization and dissociation of Naphthalene at 266, 355, and 532 nm, Journal of Photochemistry and Photobiology A: Chemistry 215 (2010) 140 – 146.

Multiphoton ionization of naphthalene clusters at 266 nm, Revista Mexicana De Fisica, 57 (2011) 68 – 74.

Multiphoton dissociation and vibrational mediated dissociation of chlorinated methanes, CH4-nCln (n=2, 3, 4) at 355nm, Journal of Photochemistry and Photobiology A: Chemistry 220 (2011) 31-38.

Asphaltene surface erosion in air plasma, Plasma Science and Technology, 12 (2010) Electron capture cross sections for Li+ colliding on H and H2 in the low keV energy region, Physical Review A, 82 (2010) 1-7.

The Effect of the Argon Carrier Gas in the Multiphoton Dissociation-Ionization of Tetracene, International Journal of Molecular Sciences, 9 (2008) 2003 – 2015.

Photoionization and dissociation pathways for naphthalene at 266 nm, Revista Mexicana De Fisica, 55 (2009) 312 – 320.

Multiphoton ionization of naphthalene clusters at 266 nm, Revista Mexicana de Física, 57 (2011) 68-74 Two, three and four photon absorption of naphthalene, Revista Mexicana de Física, 55 (2009) 312-320.

In: Chemistry Researcher Biographical Sketches
Editors: E. S. Huseynov and V. Babayev
ISBN: 978-1-62100-771-5

Chapter 71

BIOGRAPHICAL SKETCH

Name:	Title:	Date of Birth:
Tatiana Prado	Laboratory of Technological Development in Virology, Oswaldo Cruz Institute, Oswaldo Cruz Foundation (FIOCRUZ).	November 24th 1979

EDUCATION:

Institution and Location	Degree	Year Conferred	Scientific Field
Graduate in Biological Sciences at the Universidade Estadual de Campinas (UNICAMP – São Paulo, SP, Brazil), pos-graduate (masters degree) in Public Health at the Public Health National School – at the Oswaldo Cruz Foundation (FIOCRUZ, Rio de Janeiro, Brazil) and is currently a PhD student in Sciences at the Oswaldo Cruz Institute (IOC/FIOCRUZ, Rio de Janeiro, Brazil).			

CONTACT POINTS:

Av. Brasil - 4.365, Manguinhos, CEP 21040-360, Rio de Janeiro, RJ, Brazil

RESEARCH AND PROFESSIONAL EXPERIENCE:

For ten years has worked on issues related to environmental sanitation and public health, evaluating performance from wastewater treatment plants (anaerobic and aerobic processes) in microorganism's removal (antimicrobial resistant bacterial, acute gastroenteritis viruses and hepatitis A virus) and presence of pathogens in sewage sludge and biosolids produced in wastewater treatment plants. Recently, has developed projects on molecular epidemiology of viral strains circulating into the environment and improvement methods to detecting enteric viruses in sewage sludge and biosolids.

Professional Appointments:

PhD student (doctorate in Science) at the Oswaldo Cruz Institute of Oswaldo Cruz Foundation (FIOCRUZ), Rio de Janeiro, RJ, Brazil

Awards And Honors:

Award for best paper presented at the XXIV Brazilian Congress of Sanitary and Environmental Engineering (2007): "Profile of antimicrobial resistant bacterial detected in effluents from a hospital wastewater treatment plant". (Brazilian Association of Sanitary and Environmental Engineering – ABES).

Publications during last three years:

Prado, T., Silva, D.M., Guilayn, W.C., Rose, T.L., Gaspar, A.M.C., Miagostovich, M.P. (2011). Quantification and molecular characterization of enteric viruses detected in effluents from two hospital wastewater treatment plants. Water Res. 45(3), 1287-1297.

Prado, T., Miagostovich, M. P., Gaspar, A.M.C. (2010) Comparative analysis of four methods for recovery hepatitis A virus and gastroenteritis viruses from biosolids. In: XXI National Meeting of Virology and V Mercosur Meeting of Virology, 2010, Gramado - RS. Virus Reviews & Research. Rio de Janeiro : Journal of the Brazilian Society for Virology, 2010. v. 15. p. 240-241.

Miagostovich, M.P., Fumian, T. M., Prado, T., Guimaraes, F. R., Victoria, M., Gaspar, A. M. C., Leite, J. P. G. (2010) Detección de virus en aguas residuales en Rio de Janeiro, Brasil. In: Simpósio Latino Americano de Microbiologia, 2010, Montevidéo. Simpósio Latino Americano de Microbiologia, 2010.

In: Chemistry Researcher Biographical Sketches
Editors: E. S. Huseynov and V. Babayev
ISBN: 978-1-62100-771-5

Chapter 72

BIOGRAPHICAL SKETCH

Name:	Title:	Date of Birth:
Antonio Jose Queimada	LSRE/LCM - Laboratory of Separation and Reaction Engineering	16th September 1976

EDUCATION:

Institution and Location	Degree	Year Conferred	Scientific Field
Industrial Chemistry, Aveiro University, Portugal (1999), PhD. Chemical Engineering, Aveiro University, Portugal (2004)			

CONTACT POINTS:

Faculdade de Engenharia da Universidade do Porto, Rua Roberto Frias, 4200-465 Porto, Portugal

RESEARCH AND PROFESSIONAL EXPERIENCE:

Assistant Researcher (2006-)

Publications during last three years:

1. Mariana B. Oliveira, João A. P. Coutinho and António J. Queimada, Surface Tensions of Esters from a Combination of the Gradient Theory with the CPA EoS, accepted for publication in Fluid Phase Equilibria, 2011.

2. Fátima L. Mota, António J. Queimada, Simão P. Pinho and Eugénia A. Macedo, Solubility of Drug-Like Molecules in Pure Organic Solvents with the CPA EoS, accepted for publication in Fluid Phase Equilibria, 2011.
3. Mariana B. Oliveira, Vítor Ribeiro, António J. Queimada and João A. P. Coutinho, Modeling Phase Equilibria Relevant to Biodiesel Production: A Comparison of gE Models, Cubic EoS, EoS-gE, and Association EoS, Industrial and Engineering Chemistry Research, , 2011, 50, 2342-2358.
4. Mariana B. Oliveira, António J. Queimada, Georgios M. Kontogeorgis and João A. P. Coutinho, Evaluation of the CO2 Behavior in Binary Mixtures with Alkanes, Alcohols, Acids and Esters using the Cubic-Plus-Association Equation of State, Journal of Supercritical Fluids, 2011, 55, 876-892.
5. Fátima L. Mota, António J. Queimada, Simão P. Pinho and Eugénia A. Macedo, Water solubility of Drug-like Molecules with the Cubic-plus-Association Equation of State, Fluid Phase Equilibria, 2010, 298.75.
6. Nuno M. Garrido, Miguel Jorge, António J. Queimada, Ioannis G. Economou and Eugénia A. Macedo, Molecular Simulation of the Hydration Gibbs Energy of Barbiturates, Fluid Phase Equilibria, 2010, 289, 148.
7. Mariana B. Oliveira, António José Queimada, João A.P. Coutinho, Prediction of Near and Supercritical Fatty Acid Ester + Alcohol Systems with the CPA EoS, Journal of Supercritical Fluids, 2010, 52, 241-248.
8. Mariana B. Oliveira, Sofia I. Miguel, António J. Queimada, and João A. P. Coutinho, Phase Equilibria of Ester + Alcohol Systems and Their Description with the Cubic-Plus-Association Equation of State, Industrial & Engineering Chemistry Research, 2010, 49, 3452-3458.
9. Miguel Jorge, Nuno M. Garrido, António J. Queimada, Ioannis G. Economou and Eugénia A. Macedo, Effect of the Integration Method on the Accuracy and Computational Efficiency of Free Energy Calculations Using Thermodynamic Integration, Journal of Chemical Theory and Computation, 2010, 6, 1018.
10. Nuno M. Garrido, António J. Queimada, Miguel Jorge, Ioannis G. Economou and Eugénia A. Macedo, Molecular Simulation of Absolute Hydration Gibbs Energies of Polar Compounds, Fluid Phase Equilibria, 2010, 296, 110.
11. Mariana B. Oliveira, António J. Queimada, and João A. P. Coutinho, Modeling of Biodiesel Multicomponent Systems with the Cubic-Plus-Association (CPA) Equation of State, Industrial & Engineering Chemistry Research, 2010, 49, 1419.
12. Nuno M. Garrido, António J. Queimada, Miguel Jorge, Eugénia A. Macedo and Ioannis G. Economou, 1-Octanol/Water Partition Coefficients of n-Alkanes from Molecular Simulations of Absolute Solvation Free Energies, Journal of Chemical Theory and Computation, 2009, 5, 2436.
13. Fátima L. Mota, Aristides P. Carneiro, António J. Queimada, Simão P. Pinho and Eugénia A. Macedo, Temperature and Solvent Effects in the Solubility of Some Pharmaceutical Compounds: Measurements and Modeling, European Journal of Pharmaceutical Sciences, 2009, 37, 499.
14. Mariana B. Oliveira, Vera L. Oliveira, João A. P. Coutinho and António J. Queimada, Thermodynamic Modeling of the Aqueous Solubility of PAHs, Industrial & Engineering Chemistry Research, 2009, 48, 5530.

15. Mariana B. Oliveira, A. R. R. Teles, António J. Queimada and João A. P. Coutinho, Phase Equilibria of Glycerol Containing Systems and their Description with the Cubic-Plus-Association (CPA) Equation of State, Fluid Phase Equilibria,2009, 280, 22.
16. Mariana B. Oliveira, Maria. J. P. de Melo, Isabel M. Marrucho, António J. Queimada, João A. P. Coutinho, Description of the Mutual Solubilities of Fatty Acids and Water with the CPA EoS, AIChE J.,2009, 55, 1604.
17. António J. Queimada, Fátima L. Mota, Simão P. Pinho and Eugénia A. Macedo, Solubility of Biologically Active Phenolic Compounds: Measurements and Modeling (correction), J. Phys. Chem. B., 2009, 113, 6582.
18. António J. Queimada, Fátima L. Mota, Simão P. Pinho and Eugénia A. Macedo, Solubility of Biologically Active Phenolic Compounds: Measurements and Modeling, J. Phys. Chem. B., 2009, 113, 3469.
19. Fátima L. Mota, António J. Queimada, Simão P. Pinho and Eugénia A. Macedo, Aqueous Solubility of Some Natural Phenolic Compounds, Industrial & Engineering Chemistry Research, 2008, 47, 5182.
20. Pedro J. Carvalho, Mara G. Freire, Isabel M. Marrucho, António J. Queimada and João A. P. Coutinho, Surface tensions for the [Cnmim][Tf2N] ionic liquids, Journal of Chemical and Engineering Data, 2008, 53, 1346.
21. Mariana B. Oliveira, I.M. Marrucho, J.A.P. Coutinho and A.J. Queimada, Surface Tension of Chain Molecules Through a Combination of the Gradient Theory With the CPA EoS, Fluid Phase Equilibria, 2008, 267, 83.
22. Mariana B. Oliveira, Fátima R. Varanda, Isabel M. Marrucho, António J. Queimada and João A. P. Coutinho, Prediction of Water Solubility in Biodiesel With the CPA EoS, Industrial & Engineering Chemistry Research, 2008, 47, 4278.
23. Mariana B. Oliveira, João A. P. Coutinho and António J. Queimada, Mutual Solubilities of Hydrocarbons and Water with the CPA EoS, Fluid Phase Equilibria, 2007, 258, 58.
24. Mara G. Freire, Pedro J. Carvalho, Ana M. Fernandes, Isabel M. Marrucho, António J. Queimada and João A.P. Coutinho, Surface Tensions of Imidazolium Based Ionic Liquids: Anion, Cation, Temperature and Water Effect, Journal of Colloid and Interface Science, 2007, 314, 621.
25. Mariana B. Oliveira, Mara G. Freire, Isabel M. Marrucho, João A. P. Coutinho, Georgios M. Kontogeorgis and António J. Queimada, Modeling the Liquid-Liquid Equilibria of Water + Perfluorocarbons with the Cubic-plus-Association Equation of State, Industrial & Engineering Chemistry Research, 2007, 46, 1415.

In: Chemistry Researcher Biographical Sketches
Editors: E. S. Huseynov and V. Babayev
ISBN: 978-1-62100-771-5

Chapter 73

BIOGRAPHICAL SKETCH

Name:	Title:	Date of Birth:
Stéphane L. Raeppel	MethylGene Inc	15 December 1968, France

EDUCATION:

Institution and Location	Degree	Year Conferred	Scientific Field
Ph.D. in organic chemistry from Université Louis Pasteur de Strasbourg, France			

CONTACT POINTS:

MethylGene Inc., Department of Medicinal Chemistry, 7220 Frederick-Banting, Montréal, Qc, H4S 2A1, Canada

ChemRF Laboratories Inc., 3194, rue Claude-Jodoin, Montréal, Qc, H1Y 3M2, Canada

RESEARCH AND PROFESSIONAL EXPERIENCE:

2009-now Co-owner and VP of ChemRF Laboratories Inc.
2009-now Contract Medicinal Chemist, MethylGene Inc., Department of Medicinal Chemistry
2006-2009 Group Leader, MethylGene Inc., Department of Medicinal Chemistry
2003-2006 Sr. Research Chemist II, MethylGene Inc., Department of Medicinal Chemistry
2000-2003 Sr. Research Chemist I, MethylGene Inc., Department of Medicinal Chemistry
1999-2000 Postdoctoral Fellow, Research Supervisor: Professor Mark Lautens, University of Toronto, Toronto, Ontario, Canada

1991-1992 Technician in Organic Chemistry, Laboratoire NEOSYSTEM (now called *NeoMPS SA*), Strasbourg, France

1889-1990 Technician in Chemistry, Rohm & Haas France, Laboratoire de Recherche et Développement Dithane, Lauterbourg, France

Publications during last three years:

- N. Zhou, O. Moradei, S. Raeppel, S. Leit, S. Frechette, F. Gaudette, I. Paquin, N. Bernstein, G. Bouchain, A. Vaisburg, Z. Jin, J. Gillespie, J. Wang, M. Fournel, P. T. Yan, M.-C. Trachy-Bourget, A. Kalita, A. Lu, J. Rahil, A. R. MacLeod, Z. Li, J. M. Besterman, D. Delorme. "Discovery of *N* -(2-aminophenyl)-4-[(4-pyridin-3-ylpyrimidin-2-ylamino)methyl]benzamide (MGCD0103), an orally active histone deacetylase inhibitor." *J. Med. Chem.* 2008, *51*, 4072-4075.
- S. Raeppel, N. Zhou, F. Gaudette, S. Leit, I. Paquin, G. Larouche, O. Moradei, S. Fréchette, L. Isakovic, D. Delorme, M. Fournel, A. Kalita, A. Lu, M.-C. Trachy-Bourget, P. T. Yan, J. Liu, J. Rahil, J. Wang, J. M. Besterman, K. Murakami, Z. Li, A. Vaisburg. "SAR and biological evaluation of analogues of a small molecule histone deacetylase inhibitor *N* -(2-aminophenyl)-4-((4-(pyridin-3-yl)pyrimidin-2-ylamino)methyl) benzamide (MGCD0103)." *Bioorg. Med. Chem. Lett.* 2009, *19*, 644-649.
- S. Raeppel, S. Claridge, O. Saavedra, F. Gaudette, L. Zhan, M. Mannion, N. Zhou, F. Raeppel, M.-C. Granger, L. Isakovic, R. Déziel, H. Nguyen, N. Beaulieu, C. Beaulieu, I. Dupont, M.-F. Robert, S. Lefebvre, M. Dubay, J. Rahil, J. Wang, H. Ste-Croix, A. R. Macleod, J. M. Besterman, A. Vaisburg. " *N* -(3-Fluoro-4-(2-arylthieno[3,2-*b*]pyridin-7-yloxy)phenyl)-2-oxo-3-phenylimidazolidine-1-carboxamides: a novel series of dual c-Met/VEGFR2 receptor tyrosine kinase inhibitors." *Bioorg. Med. Chem. Lett.* 2009, *19*, 1323-1328.
- M. Mannion, S. Raeppel, S. Claridge, N. Zhou, O. Saavedra, L. Isakovic, L. Zhan, F. Gaudette, F. Raeppel, R. Déziel, N. Beaulieu, H. Nguyen, I. Chute, C. Beaulieu, I. Dupont, M.-F. Robert, S. Lefebvre, M. Dubay, J. Rahil, J. Wang, H. Ste-Croix, A. R. Macleod, J. M. Besterman, A. Vaisburg. "*N* -(4-(6,7-disubstituted-quinolin-4-yloxy)-3-fluorophenyl)-2-oxo-3-phenylimidazolidine-1-carboxamides: a novel series of dual c-Met/VEGFR2 receptor tyrosine kinase inhibitors." *Bioorg. Med. Chem. Lett.* 2009, *19*, 6552-6556.
- O. Saavedra, S. Claridge, L. Zhan, F. Raeppel, M.-C. Granger, S. Raeppel, M. Mannion, F. Gaudette, N. Zhou, L. Isakovic, N. Bernstein, R. Déziel, H. Nguyen, N. Beaulieu, C. Beaulieu, I. Dupont, J. Wang, A. R. Macleod , J. M. Besterman, A. Vaisburg. "N^3 -Arylmolonamides: A new series of thieno[3,2-*b*]pyridine based inhibitors of c-Met and VEGF2 tyrosine kinases." *Bioorg. Med. Chem. Lett.* 2009, *19*, 6836-6839.
- F. Gaudette, S. Raeppel, H. Nguyen, N. Beaulieu, C. Beaulieu, I. Dupont, A. R. Macleod, J. M. Besterman, A. Vaisburg. "Identification of potent and selective VEGFR receptor tyrosine kinase inhibitors having new amide isostere headgroups." *Bioorg. Med. Chem. Lett.* 2010, *20*, 848-852.

- S. Raeppel, F. Gaudette, M. Mannion, S. Claridge, O. Saavedra, L. Isakovic, R. Déziel, N. Beaulieu, C. Beaulieu, I. Dupont, H. Nguyen, J. Wang, A. R. Macleod, C. Maroun, J. M. Besterman, A. Vaisburg . "Identification of a novel series of potent RON receptor tyrosine kinase inhibitors." *Bioorg. Med. Chem. Lett.* 2010, *20*, 2745-2749.

In: Chemistry Researcher Biographical Sketches
Editors: E. S. Huseynov and V. Babayev
ISBN: 978-1-62100-771-5

Chapter 74

BIOGRAPHICAL SKETCH

Name:	Title:	Date of Birth:
Michael A. Reshchikov	Physics Department, Virginia Commonwealth University	June 11, 1959

EDUCATION:

Institution and Location	Degree	Year Conferred	Scientific Field
PhD in Physics/Mathematics from Ioffe Physical Technical Institute (St.-Petersburg, Russia)			

CONTACT POINTS:

701 West Grace St., Richmond, VA 23238

RESEARCH AND PROFESSIONAL EXPERIENCE:

Expert in photoluminescence from defects in semiconductors. Author and co-author of more than 100 journal papers, including the review paper "Luminescence properties of defects in GaN published in J. Appl. Phys 97, 061301 (2005) and cited more than 300 times.

Professional Appointments:

Associate Professor at Physics Department of VCU

Publications during last three years:

1. M. A. Reshchikov, A. El-Shaer, A. Behrends, A. Bakin, and A. Waag, *On quantum efficiency of photoluminescence in ZnO layers and nanostructures*, Physica B **404**, 4813 (2009).
2. M. A. Reshchikov, M. Foussekis, and A. A. Baski, *Surface photovoltage in undoped n-type GaN*, J Appl. Phys. **107**, 113535 (2010).
3. M. A. Reshchikov, A. Kvasov, T. McMullen, M. F. Bishop, A. Usikov, V. Soukhoveev, and V. A. Dmitriev, *Dramatic thermal quenching of photoluminescence in high-resistivity Zn-doped GaN*, accepted to Phys. Rev. B on 06/09/2011 (ms # LQ12477B).

In: Chemistry Researcher Biographical Sketches
Editors: E. S. Huseynov and V. Babayev

ISBN: 978-1-62100-771-5

Chapter 75

BIOGRAPHICAL SKETCH

Name:	Title:	Date of Birth:
Claudionor Ribeiro da Silva	Federal University of Uberlândia	October 29, 1975

EDUCATION:

Institution and Location	Degree	Year Conferred	Scientific Field
Graduation in Surveying Engineering, Federal		2004	
University of Viçosa - Minas Gerais / Brazil; M.Sc in Geodesy,		2006	
Federal University of Paraná Paraná / Brazil; and Ph.D. in Geodesy,		2008	
Federal University of Paraná Paraná / Brazil;			

CONTACT POINTS:

Avenue Goiás, n. 2000, Vila Nova, ZipCode: 38500-000, Monte Carmelo, Minas Gerais / Brazil.

RESEARCH AND PROFESSIONAL EXPERIENCE:

Professor at Federal University of Piauí (five years); Piauí/Brazil. Professor at Federal University of Piauí (one month); Piauí/Brazil. Coordinator of four projects ofresearches at Federal University of Piauí.

Professional Appointments:

Research and Professor at Federal
University of Uberlândia

Publications during last three years:

10 papers in congresses, 7 full papers and 1 abstract in scientific journals and 2 book chapter.

In: Chemistry Researcher Biographical Sketches
Editors: E. S. Huseynov and V. Babayev

ISBN: 978-1-62100-771-5

Chapter 76

BIOGRAPHICAL SKETCH

Name:	Title:	Date of Birth:
Semenov Semen	Institute of Biochemical Physics RAS, Russia	02.28.1951

EDUCATION:

Institution and Location	Degree	Year Conferred	Scientific Field
Moscow Steel and Alloys Institute			

CONTACT POINTS:

Institute of Biological Physics, Russian Academy of Science, Kosygin Street 4, Russia.

Research and Professional Experience: Physicochemical hydrodynamics in application to analytical science, Theory of mass transport in liquids.

Professional Appointments:

Leading researcher at Institute of Biochemical Physics RAS, Cand. Sci. (Physical Chemistry), Moscow State University, 1987.

Publications during last three years:

- **Mass transport thermodynamics in nonisothermal molecular liquid mixtures** S. N. Semenov, M E Schimpf, Physsics-Uspekhi, Volume 52 (2009), Number 10, pp. 1045–1054
- **Statistical thermodynamic expression for the Soret coefficient** S. N. Semenov, EPL (Europhysics Letters) Volume 90 Number 5, 2010, 56002

- **Theory of strong temperature dependence in thermophoresis** S. N. Semenov, in Thermal Nonequilibium - Lecture Notes of the 8th International Meeting on Thermodiffusion, Editors: S. Wiegand, W. Köhler, J. K. G. Dhont, 2008, pp. 117-124
- **Re-evaluation of the thermodynamics approach to mass transport in temperature and concentration gradients** S. N. Semenov, in Thermal Nonequilibium - Lecture Notes of the 8th International Meeting on Thermodiffusion, Editors: S. Wiegand, W. Köhler, J. K. G. Dhont, 2008, pp. 109-117
- **Thermodynamics of mass transport in diluted suspensions of charged particles in non-isothermal liquid electrolytes** Semen N. Semenov and Martin E. Schimpf, Comptes Rendus Mécanique, Volume 339, Issue 5, 2011, pp. 280-286
- **Internal degrees of freedom, molecular symmetry and thermodiffusion** Semen N. Semenov and Martin E. Schimpf, Comptes Rendus Mécanique, Volume 339, Issue 5, 2011, pp. 335-341
- **Phase Layering in Binary Liquid Mixtures Due to Thermodiffusion** Semen N. Semenov and Martin E. Schimpf, in Thermal Nonequilibium - Lecture Notes of the 8th International Meeting on Thermodiffusion, Editors: S. Wiegand, W. Köhler, J. K. G. Dhont, 2008, pp. 63-71

In: Chemistry Researcher Biographical Sketches
Editors: E. S. Huseynov and V. Babayev
ISBN: 978-1-62100-771-5

Chapter 77

BIOGRAPHICAL SKETCH

Name:	Title:	Date of Birth:
Konstantin Zh.Seminsky	Institute of the Earth's Crust (Siberian Branch of Russian Academy of Sciences)	09/10/1961

EDUCATION:

Institution and Location	Degree	Year Conferred	Scientific Field
geologist, Irkutsk State Technical University (1983)			

CONTACT POINTS:

128, Lermontov Street, Irkutsk, 664033, Russia
Research and Professional Experience:

Research and Professional Experience:

Structural geology, tectonophysics, soil–gas survey

Professional Appointments:

Head of Laboratory, Doctor of Geology and Mineralogy

Publications during last three years:

Seminskii K.Zh. 2008. Hierarchy in the Zone-Block Lithospheric Structure of Central and Eastern Asia // Russian Geology and Geophysics. V. 49. P. 771–779.

Seminsky K.Zh., Gladkov A.S., Vakchromeev A.G. et al. 2008. Faults and Seismicity of the South of Siberian Platform: Feature of Display at Different Scale Levels // Lithosphere. V. 4. P. 3-21 (in Russian).

Seminsky K.Zh., Gladkov A.S., Radziminovich Ya.B. et al. 2008. Regularities of Manifestation of Active Faults and Seismicity in the Southern Part of the Siberian Craton // Doklady Earth Sciences. V. 422, No. 7. P. 1068–1072.

Seminsky K.Zh. 2009. Major Factors of the Evolution of Basins and Faults in the Baikal Rift Zone: Tectonophysical Analysis // Geotectonics. V. 43, No. 6. P. 486–500.

Seminsky K.Zh., Bobrov A.A. 2009. Radon Activity of Faults (Western Baikal and Southern Angara Areas) // Russian Geology and Geophysics. V. 50. P. 682–692.

Seminskii K.Zh., Bobrov A.A. 2009. Comparative Assessment of Radon Activity for Different Fault Types and Scale Ranks in the Baikal Rift and South of the Siberian Platform // Doklady Earth Sciences. V. 427A, No. 6. P. 915–919.

Seminsky K.Zh., Kogut E.I. 2009. Governing Factors in the Development of Depressions and Faults in the Baikal Rift Zone: Results of a Physical Experiment // Doklady Earth Sciences. V. 424, No. 1. P. 15–18.

Sherman S.I., Seminsky K.Zh. 2010. Tectonophysical Research at Institute of the Earth's Crust SB RAS: Major Achievements and Actual Problems // Geodynamics & Tectonophysics. V. 1. No 1. P. 4–23 (http://www.crust).

Seminsky K.Zh., Cheremnykh A.V. 2011. Jointing Patterns and Stress Tensors in Cenozoic Sediments of the Baikal Rift: Development of the Structural-Genetic Approach // Russian Geology and Geophysics. V. 52. P. 353–367.

Seminsky K.Zh., Radziminovich Ya.B. 2011. The Cross Sizes and Lateral Zoning of the Baikal Seismic Belt // Doklady Earth Sciences. V. 438, No. 1.

Additionally for last 3 years 12 articles were published by Seminsky K.Zh. in Proceedings of different conferences (in Russian and English).

In: Chemistry Researcher Biographical Sketches
Editors: E. S. Huseynov and V. Babayev
ISBN: 978-1-62100-771-5

Chapter 78

BIOGRAPHICAL SKETCH

Name:	Title:	Date of Birth:
Mohd. Shakir Khan	Aligarh Muslim University, Aligarh (U.P), India	20th September, 1979

EDUCATION:

Institution and Location	Degree	Year Conferred	Scientific Field
(Under the process of Submission) from Aligarh Muslim University, Aligarh	Phil & Ph. D	2007	

CONTACT POINTS:

Department of Applied Physics, Z. H. College of Engineering & Technology, Aligarh Muslim University, Aligarh-202002 (India)

RESEARCH AND PROFESSIONAL EXPERIENCE:

About 6 Years

Professional Appointments:

Research Scholar since 20th October, 2005 to till date (at Aligarh Muslim University, Aligarh)

Awards and Honors:

1. UGC fellowship during 01/01/2007 to 19/10/2007
2. UGC fellowship during 29/12/2007 to till
3. ICTP grant to attend the International workshop on Nuclear Sciences and its Peaceful applications (NSPA) at Faculty of Science, Cairo University, Egypt, 2009.

Publications during last three years:

1. M. Shakir Khan, A. H. Naqvi and Ameer Azam (*2008*): Study of Indoor Radon and its Progeny Levels in Rural Areas of North India Using LR-115 Plastic Track detectors. Radiation Measurements; *Vol. 43, S385–S388.*
2. K. Mahur, M. Shakir Khan, A. H. Naqvi, Rajendra Prasad and Ameer Azam (*2008*): Measurement of effective Radium Content of sand samples collected from Chhatrapur Beach, Orissa, India using track etch technique. Radiation Measurements; *Vol. 43, S520– S522.*
3. M. Shakir Khan, A. H. Naqvi and Ameer Azam *(2009)*: Depth Dependence Radon Study in Indian Tube-Wells. Asian Journal of Chemistry; *Vol. 21, No. 10, S292*–295.
4. M. Shakir Khan, Ameer Azam, A.H. Naqvi, Deepak Verma, M. Zubair and M.K. Bhardwaj *(2010)*: Radium and Radon Exhalation Studies in Soil Samples. Recent Trends in Radiation Physics Research. *ISBN: 978-81-7906-227-2. pp. 356–357.*
5. M. Shakir Khan, M. Zubair, Deepak Verma, A.H. Naqvi, Ameer Azam and M.K. Bhardwaj *(2010)*: The Study of Indoor Radon in the Urban Dwellings Using Plastic Track Detectors. Environmental Earth Sciences *(DOI: 10.1007/s12665-010-0701-5).*
6. M. Shakir Khan, A.H. Naqvi, A. Azam and D.S. Srivastava *(2011)*: Radium and Radon Exhalation Studies of Soil. Iranian journal of Radiation Research, 8(4)x.

In: Chemistry Researcher Biographical Sketches
Editors: E. S. Huseynov and V. Babayev
ISBN: 978-1-62100-771-5

Chapter 79

BIOGRAPHICAL SKETCH

Name:	Title:	Date of Birth:
Jose Angel Siles Lopez	Mr, Dr.	15th October, 1981

EDUCATION:

Institution and Location	Degree	Year Conferred	Scientific Field
Chemical Engineering	PhD	June 2010	
Workplace Health and Safety	Master	Febr. 2007	
Certificate of Advanced Studies on Chemical Engineering		Oct. 2005. 2006:	
	Bachelor Thesis.	Oct. 2005	
Environmental Sciences.	Bachelor	Oct. 1999 June 2004	

CONTACT POINTS:

Jose Angel Siles Lopez

RESEARCH AND PROFESSIONAL EXPERIENCE:

Biodiesel manufacturing from used cooking oils by alkaline transesterification.

Anaerobic digestion of wastewater and glycerol derived from biodiesel manufacturing.

Biomethanization of orange peel waste and wastewater derived from the pressing of orange peel generated in the orange juice production.

Anaerobic bioprocessing of vinasse and maize as energy crop. Inhibition of termophilic anaerobic digestion.

Kinetic modeling of biological and chemical processes. Squalene purification from olive oil and shark liver.

Composting of urban waste. Succinic acid production from agricultural wastes by Fibrobacter sp.

Advanced oxidation processes for wastewater treatment. Physical-chemical treatments of oily wastewater.

Professional Appointments:

Assistant Professor

Awards and Honors:

Five awards on highly polluting wastes management

Publications during last three years:

1) Siles, J.A., Martín, M.A., Martín, A., Raposo, F., Borja, R., 2007. Anaerobic digestion of wastewater derived from the pressing of orange peel generated in the orange juice production. J. Agr. Food Chem., 55(6), 1905-1914.\
2) Siles, J.A., Chica, A., Borja, R., Martín, M.A., 2007. Estudio y modelización de la digestión anaerobia de las aguas residuales del prensado de la cáscara de naranja. Tecnología del Agua, 27(291), 44-60.
3) Berrios, M., Siles, J.A, Martín, M.A., Martín, A., 2007. A kinetic study of the esterification of free fatty acids (FFA) in sunflower oil. Fuel, 86, 2383-2388.
4) Siles, J.A., Martín, M.A., Chica, A.F., Borja, R., 2008. Kinetic modelling of the anaerobic digestion of wastewater derived from the pressing of orange rind produced in orange juice manufacturing. Chem. Eng. J., 140, 145-156.
5) Chica, A.F., Martín, M.A., Siles, J.A., Dios, M., 2008. La medida de la actividad respiratoria durante el compostaje de residuos sólidos municipales. Reciclado y plásticos, 29, 7-8.
6) Siles, J.A., Martín, M. A., Chica, A.F., Martín, A., 2009. Anaerobic digestion of glycerol derived from biodiesel manufacturing. Bioresource Technol., 100(23), 5609-5615.
7) Siles, J.A., Li, Q., Thompson, I., 2010. Biorefinery of waste orange peel. Crit. Rev. Biotechnol. 30(1), 63-69.
8) Siles, J.A., Martín, M.A., Chica, A.F., Martín, A., 2010. Anaerobic co-digestion of glycerol and wastewater derived from biodiesel manufacturing. Bioresource Technol., 101, 6315-6321.\

9) Martín, M.A., Siles, J.A., Chica, A.F., Martín, A., 2010. Modelling the anaerobic digestion of wastewater derived from the pressing of orange peel produced in orange juice manufacturing. Bioresource Technol., 101, 3909?3916.
10) Siles, J.A., Brekelmans, J., Martín, M.A., Chica, A.F., Martín, A. 2010. Impact of ammonia and sulphate concentration on thermophilic anaerobic digestion. Bioresource Technol., 101, 9040-9048.
11) Martín, M.A., Siles, J.A., Chica, A.F., Martín, A. 2010. Biomethanization of orange peel waste. Bioresource Technol., 101, 8993-8999.
12) Li, Q., Siles, J.A., Thompson, I. 2010. Succinic acid production from orange peel and wheat straw by batch fermentations of Fibrobacter succinogenes S85. Appl. Microbiol. Biot. 88(3), 671-678.

Patents:

1) Aguado, M., Chica, A., Berrios, M., Siles, J.A., Martín, M.A., Martín, A. A method to obtain squalene from animal origin. N° P200700462. Date: 22nd of Febrary, 2007.
2) Aguado, M., Chica, A., Berrios, M., Siles, J.A., Martín, M.A., Martín, A. A method to obtain squalene from vegetable origin. N°P200701369. Date: 18th of May, 2007.

In: Chemistry Researcher Biographical Sketches
Editors: E. S. Huseynov and V. Babayev
ISBN: 978-1-62100-771-5

Chapter 80

BIOGRAPHICAL SKETCH

Name:	Title:	Date of Birth:
Steven Y. Qian	Ph.D	09/20/1960

EDUCATION:

Institution and Location	Degree	Year Conferred	Scientific Field
Department of Pharmaceutical Sciences, College of Pharmacy, Nursing and Allied Sciences, North Dakota State University, Fargo, ND 58108			

CONTACT POINTS:

North Dakota State University
NDSU Dept 2665
208B Sudro Hall, P.O. Box 6050
Fargo, ND 58108-6050

RESEARCH AND PROFESSIONAL EXPERIENCE:

1994-1999 Graduate Research Assistant, Free Radical and Radiation Biology Program, College of Medicine, The University of Iowa, Iowa City, IA.

1999-2004 Post-doctoral/IRTA Research Fellow, Laboratory of Pharmacology and Chemistry, National Institute of Environmental Health Sciences (NIEHS), Research Triangle Park, NC.

2004-2005 Research Assistant Professor, Division of Drug Delivery and Disposition, School of Pharmacy, The University of North Carolina at Chapel Hill, NC.

2004-2007 Guest Researcher, Laboratory of Pharmacology and Chemistry, NIEHS, Research Triangle Park, NC.

2005-date Assistant Professor, Department of Pharmaceutical Sciences, College of Pharmacy, North Dakota State University, Fargo, ND.
2006-date Assistant Professor/Faculty Member, Department of Cellular and Molecular Biology, North Dakota State University, Fargo, ND.
2010-date Assistant Professor/Faculty Member/Researcher, Center for Protease Research, North Dakota State University, Fargo, ND

Professional Appointments:

Assistant Professor @ NDSU

Awards and Honors:

1996 Young Investigator Award, Society for Free Radical Biology & Medicine (SFRBM)
1996 Travel Award, Radiation Research Society
1997 Travel Award, American Society for Photobiology
1998 Travel Award, American Society for Photobiology
2001 Young Investigator Award, Society for Free Radical Biology & Medicine (SFRBM)
2002 Young Investigator Award, International Symposium of EPR Spin Trapping
2003 Research Excellence Award (FARE), National Institutes of Health

In: Chemistry Researcher Biographical Sketches
Editors: E. S. Huseynov and V. Babayev

ISBN: 978-1-62100-771-5

Chapter 81

BIOGRAPHICAL SKETCH

Name:	Title:	Date of Birth:
Y. Tai Clifford	National Taiwan University	February 13, 1946

EDUCATION:

Institution and Location	Degree	Year Conferred	Scientific Field
Chemical Engineering, North Carolina State University, U.S.A.	Ph. D.		

CONTACT POINTS:

Department of Chemical Engineering, National Taiwan University, No. 1, Sec. 4, Roosevelt Road, Taipei, Taiwan, 106

AWARDS AND HONORS:

Consulting Editor, Journal of the Taiwan Institute of Chemical Engineers

RESEARCH AND PROFESSIONAL EXPERIENCE:

Crystallization, Nanoparticle Formation Process, Supercritical Fluid Technology

Professional Appointments:

Editor-in-Chief, Journal of Chinese Institute of Chemical Engineers.(2000-2006)

Publications during last three years

C. Y. Tai, Y.-H. Wang and H.-S. Liu, "A Green Process for Preparing Silver Nanoparticles Using Spinning Disk Reactor," *AIChE J.*, 54, 445 (2008) (SCI, EI)

C. Y. Tai, C.-K. Wu, and M.-C. Chang, "Effect of Magnetic Field on the Crystallization of $CaCO_3$ Using Permanent Magnets," *Chem. Eng. Sci.*, 63, 5606 (2008). (SCI, EI)

C. Y. Tai, and C.-K. Chen, "Particle Morphology, Habit and Size Control of $CaCO_3$ Using Reverse Microemulsion Technique," *Chem. Eng. Sci.*, 63, 3632 (2008). (SCI, EI)

C. Y. Tai, M.-C. Chang, R.-J. Shieh, and Ted. G. Chen, "Magnetic Effects on Crystal Growth Rates of Calcite in a Constant-Composition Environment," *J. Crystal Growth*, 310, 3690 (2008). (SCI, EI)

Y.-S. Chen, Y.-C. Hsu, C.-C. Lin, C. Y. Tai, and H.-S. Liu, "Volatile Organic Compounds Absorption in a Cross-Flow Rotating Packed Bed," *Environ. Sci. Technol.*, 42, 2631 (2008). (SCI, EI)

T. Y. Wang, Y.-C. Lin, C. Y. Tai, R. Sivakumar, D. K. Rai, and C. W. Lan, "A Novel Approach for Recycling of Kerf Loss Silicon from Cutting Slurry Waste for Solar Cell Application," *J. Crystal, Growth*, 310, 3403 (2008). (SCI, EI).

C. Y. Tai, Y.-H. Wang, Y. W. Kuo, M. H. Chang, and H. S. Liu, "Synthesis of Silver Particles below 10 nm Using Spinning Disk Reactor," *Chem. Eng. Sci.*, 64, 3112 (2009). (SCI, EI)

C. Y. Tai, Y. H. Wang, C.-T. Tai, and H.-S. Liu, "Preparation of Silver Nanoparticles Using a Spinning Disk Reactor in a Continuous Mode," *I&EC Res.*, 48, 10104 (2009). (SCI, EI)

C. Y. Tai, C.-D. Tai, and M.-H. Chang, "Effect of Interfacial Supersaturation on Secondary Nucleation," *J. Taiwan Inst. Chem. Engrs.*, 40, 439 (2009). (SCI, EI)

C. Y. Tai, "Experimental Techniques for Indentifying Operating Variables of Crystal Growth and Nucleation—From Readily-Soluble Salts to Sparingly-Soluble Salts," *J. Taiwan Inst. Chem. Engrs.*, 40, 682 (2009). (SCI, EI)

Y.-C. Lin and C. Y. Tai, "Recovery of Silicon Powder from Kerfs Loss Slurry Using Phase-Transfer Separation Method," *Sep. Purif. Technol.*, 74, 170 (2010). (SCI, EI)

Y.-C. Lin, T.-Y. Wang, C.-W. Lan, and C. Y. Tai, "Recovery of Silicon Powder from Kerfs Loss Slurry by Centrifugation," *Powder Technol.*, 200, 216 (2010). (SCI, EI)

Y.-S. Chen, Y.-H. Wang, H.-S. Liu, K.-Y. Hsu, and C. Y. Tai, "Micronization of p-Aminosalicylic Acid Particles Using High-Gravity Technique," *I&EC Res.*, 49, 8832 (2010). (SCI, EI)

M.-C. Chang and C. Y. Tai, "Effect of the Magnetic Field on the Growth Rate of Aragonite and the Precipitation of $CaCO_3$," *Chem. Eng. J.*, 164, 1 (2010). (SCI, EI)

C.-K. Chen and C. Y. Tai, "Competing Effects of Operating Variables in the Synthesis of $CaCO_3$ Particles Using the Reverse Microemulsion Technique," *Chem. Eng. Sci.*, 65, 4761 (2010). (SCI, EI)

S.-S. Wang, J.-T. Lai, M.-S. Huang, C. Y. Tai, and H.-S. Liu, "Deactivation of Isoamylase and β–Amylase in the Agitated Reactor under Supercritical Carbon Dioxide," *Bioproc. Biosyst. Eng.*, 33, 107 (2010). (SCI, EI)

J.-H. Yang, S.-H. Shih, C.-I. Wu, and C. Y. Tai, "Preparation of High Surface Area $CaCO_3$ for SO_2 Removal by Absorption of CO_2 in Aqueous Suspensions of $Ca(OH)_2$," *Powder Technol.*, 202, 101 (2010). (SCI, EI)

C. Y. Tai, M. J. Hon, P.-C. Chen, "A Comparison of Growth Behavior between $CaCO_3$ and CaF_2 Crystals in a Constant-Composition Environment," *J. Taiwan. Inst. Chem. Engrs.*, In Press (2011). (SCI, EI)

M.-H. Chang, H.-S. Liu, and C. Y. Tai, "The Preparation of Copper Oxide Nanoparticles and Its Application in Nanofluid," *Powder Technol.*, 207, 378 (2011). (SCI, EI)

C. Y. Tai, M.-C. Chang, and S.-W. Yeh, "Synergetic Effect of Temperature and Magnetic Field on the Aragonite and Calcite Growth," *Chem. Eng. Sci.* 66, 1246 (2011). (SCI, EI)

C. Y. Tai, C.-Y. Su, Y.-T. Chen, S.-M. Shih, and W.-C. Chien, "Carbonation of Natural Rock and Steel Slag Using Supercritical CO_2," 11th European Meeting on Supercritical Fluids, Barcelona, Spain (2008).

C. Y. Tai, "Preparation of Ultrafine Particles Using High-Gravity Technique", International Crystallization Workshop, Seoul, Korea (2008).

Y.-C. Lin, C. Y. Tai, "Recovery of Silicon Powder From Kerf Loss Slurry by Centrifugation," 20th International Symposium on Transport Phenomena, Victoria, British Columbia, Canada (2009).

C. Y. Tai, Y.-H. Wang and Y.-W. Kuo, "A Green Process for Preparing Silver Nanoparticle below 10 nm," 2nd International Congress on Green Process Engineering, Venice, Italy (2009).

M.-C. Chang, S.-W. Yeh, C.-C. Lin, and C. Y. Tai, "Effects of Solution Condition, Magnetic Field, and Temperature on the $CaCO_3$ Crystal Growth," The 13th Asia Pacific Confederation of Chemical Engineering Congress, Taipei, Taiwan (2010).

C.-C. Li, Y.-H. Wang, and C. Y. Tai, "Preparation of AgI Nanoparticles Using a Spinning Disk Reactor in a Continuous Mode," International Seminar on Fundamental & Application of Chemical Engineering, Bali, Indonesia (2010).

M.-H. Chang and C.-Y. Tai, "Preparation and Application of CuO Nanoparticles," TechConnect World-Nanotech Conference&Expo, Anaheim CA, U.S.A. (2010).

In: Chemistry Researcher Biographical Sketches
Editors: E. S. Huseynov and V. Babayev
ISBN: 978-1-62100-771-5

Chapter 82

BIOGRAPHICAL SKETCH

Name:	Title:	Date of Birth:
Tomoki Takahashi	National Agriculture Research Organization, Japan	31th August, 1969

EDUCATION:

Institution and Location	Degree	Year Conferred	Scientific Field
Bachelor of agriculture: Tokyo University of Agriculture and Technology		1992	
Master of Agriculture, Soil Science: Tohoku University		1994	

CONTACT POINTS:

1-3-1, Kan-nondai, Tsukuba city, Ibaraki. Japan

RESEARCH AND PROFESSIONAL EXPERIENCE:

2009-present: Nutrition cycle and management in Japan
4/2005-3/2009: Nitrate removal in paddy field
10/2001-3/2005: soybean production in upland field converted from rice paddy
10/1994-9/2001: Soil characteristics of upland field converted from rice paddy

Professional Appointments:

Senior researcher, National Agriculture Research Organization, Japan, April 2010-present

Senior researcher, Prefectural Research Institute of Agriculture and Forestry, April 2005-March 2009.

Researcher, National Agriculture Research Center, Japan, October 1994- March 2005.

Publications during last three years:

Takahashi, T., Inagaki, H, Fukushima, T, Oishi, T and Matsuno, K. (2010) Increasing nitrate removal at low temperatures by incorporating organic matter into paddy fields, Soil Sci. Plant Nutri., 56, 163-167.

Takahashi, T. (2010) Future Studies for production of vegetables for commercial and processing uses, Bull. Researches and Analysis report of NARO, 9, 97-125. (In Japanese)

Takahashi T. (2009) Studies on rice terrace in Shizuoka Prefecture: Its working and benefaction, ed. Prefectural Research Institute of Agriculture and Forestry, The Shizuoka Shimbun, Shizuoka、p34-44, ISBN978-4783805465. (In Japanese)

Takahashi T. (2009) New implementation for water purification in upland tea field – rice paddies system., Jpn. J. Crop Sci.、78 (Extra 2), 318-319. (In Japanese)

Takahashi, T., Nira, R., Miyaji, N., Toda, H., Muranaka, Y., Hirono, Y., Atsumi, K., Fukushima, T., and Sugiura, S. (2009) Nitrate Concentration Trends in Streams, Springs, and Wells, From Catchment Areas Located on Fields Using Reduced Fertilizer Application, Bull. Shiozuka Res. Int. Agr. Forest., 2, 17-25. (In Japanese)

Takahashi, T. Yokoyama, R. Hayashi, S., and Iwata, I. (2008) Anion-exchange propertiy of the functional charcoal prepared from plant wastes, 5th International conference interfaces against pollution 2008, 169.

Takahashi, T., Hosokawa, H. and Matsuzaki, M. (2008) Effect of soil compression on the soil drying in seedbed and imbibitions of soybean seeds in heavy clyey upland field converted from rice paddy.. Jpn. J. Soil Sci. Plant Nutri., 79, 1-8. (In Japanese with English summary)

Takahashi, T. (2008) Estimation of nitrate removal in bare paddy fields during winter season, Kagaku to Nogyo, 6, 1-5. (In Japanese)

Takahashi, T. (2008) Nitrate removal ability in coastal sandy fields, JA Shizuoka Keizairen, 1131, 10-11. (In Japanese)

Inagaki, H., Matsuno, K., Oishi, T., Takahashi, T., Yamamoto, T., and Kurita, E. (2010) Influence of agricultural experience on the image of agriculture among children. A case study of the rurban area in Fujieda, Shizuoka, J. Rural Planning, 29, 161-166. J. Rural Planning, 28, 267-262. (In Japanese)

Inagaki, H, Matshno, K., Oishi, T., and Takahashi, T. (2010) Effect of mowing ridges on the population density of wolf speeders in rice terraces. A case study of the rice terraces of Kurasawa distinct, Kikugawa, Shizuoka. J. Rural Planning, 28, 267-272. (In Japanese)

Matsuno, K, Inagaki, H, Oishi, T, and Takahashi, T. (2010) Occurrence of rice pests in paddy fields used for carrying out environmental education activities, J. Rural Planning, 28, 261-266. (In Japanese)

Ohishi, T., Inagaki, H., Takahashi, T., Matsuno, K., Yamamoto, T., Kurita, H. (2009) Comparison of preferred characteristics of agricultural landscape in Shizuka prefecture between attributes, (2009) Landscape Res., Jpn., 72, 889-892. (In Japanese with English summary)

Ohishi, T., Inagaki, H., Takahashi, T., and Matsuno, K. (2009) Community development by using nature and landscape on rice terraces, Jpn. Regional Policy, 7, 321-328. (In Japanese with English summary)

Fukushima, T., Takahashi, T, Sugiura, S., Atsumi, K., Nira, R. (2008) Nitrate leaching from coastal sandy fields, Sand Dune Res., 55, 95-98. (In Japanese)

In: Chemistry Researcher Biographical Sketches
Editors: E. S. Huseynov and V. Babayev

ISBN: 978-1-62100-771-5

Chapter 83

BIOGRAPHICAL SKETCH

Name:	Title:	Date of Birth:
Shinji Tokonami	Ph.D., Engineering	November 2nd, 1964

EDUCATION:

Institution and Location	Degree	Year Conferred	Scientific Field
Department of Radiation Physics, Institute of Radiation Emergency Medicine, Hirosaki University	Professor		

CONTACT POINTS:

66-1 Hon-cho, Hirosaki, Aomori 036-8564

RESEARCH AND PROFESSIONAL EXPERIENCE:

Radiation measurement, dosimetry

Professional Appointments: Task group on ICRU (Uncertainty of radon measurement), Project Leader of IEC 61577-2 (International Standard), Project Leader of ISO 16641

<u>Professional Appointments:</u>

Japan Health Physics Society Award(2005)

Publications during last three years:

1. Masahiro Hosoda, Tetsuo Ishikawa, Atsuyuki Sorimachi, Shinji Tokonami, Shigeo Uchida: Development and application of continuous measurement system of radon exhalation rate, *Rev. Sci Instrum.* 82(1), 015101-1~015101-4 (2011).
2. Nabil M. Hassan, Tetsuo Ishikawa, Masahiro Hosoda, Kazuki Iwaoka, Atsuyuki Sorimachi, Sarata K. Sahoo, Miroslaw Janik, Chutima Kranrod, Hidenori Yonehara, Masahiro Fukushi, Shinji Tokonami: The effect of water content on the radon emanation coefficient for some building materials used in Japan, *Radiat. Meas.*, *46*, 232-237 (2011).
3. Yumi Yasuoka, Atsuyuki Sorimachi, Tetsuo Ishikawa, Masahiro Hosoda, Tokonami, S., Nobutoshi Fukuhori, Miroslaw Janik: Separately measuring 222Rn and 220Rn concentratons exhaled from soil using alphaGUARD and liquid scintillation methods, Radiation Protection Dosimetry (2010).
4. C. K. M. Ng, V. W. Y. Choi, Miroslaw Janik, Atsuyuki Sorimachi, Kranrod Chutima, D. Nikezic, Tokonami, S.: Long-term determination of airborne radon progeny concentrations using LR 115 detectors and the effects of thoron, Radiation Protection Dosimetry (2010).
5. Ramola Rakesh, G. Prasad, G. S. Gusain, B. S. Rautela, V. M. Choubey, D.V. Sagar, Tokonami, S., Atsuyuki Sorimachi, Sahoo Sarata Kumar, Tetsuo Ishikawa: Preliminary indoor thoron measurements in high radiation background area of southeastern coastal Orissa, India, Radiation Protection Dosimetry (2010).
6. Norbert Kavasi, Vigh Tamas, Atsuyuki Sorimachi, Tetsuo Ishikawa, Tokonami, S., Masahiro Hosoda: Effective dose to miners due to radon an thoron in a manganese mine in Hungary, Radiation Protection Dosimetry (2010).
7. Kranrod Chutima, Atsuyuki Sorimachi, Tokonami, S., Tetsuo Ishikawa, Chanyotha Supitcha, Chankow Nares: Comparative dosimetry of radon and thoron, Radiation Protection Dosimetry (2010).
8. Masahiro Hosoda, Tokonami, S., Atsuyuki Sorimachi, Tetsuo Ishikawa, Sahoo Sarata Kumar, Masahide Furukawa, Yoshitaka Shiroma, Yumi Yasuoka, Janik Miroslaw, Norbert Kavasi, Shigeo Uchida, Michikuni Shimo: Influence of soil environmental parameters on thoron exhalation rate, Radiation Protection Dosimetry (2010).
9. Sahoo Sarata Kumar, Tetsuo Ishikawa, Tokonami, S., Atsuyuki Sorimachi, Janik Miroslaw, Mohamed Hassan Nabil Mohamed, Masahiro Hosoda, Kranrod Chutima, Chanyotha Supitcha, Vangeline Parami, Hidenori Yonehara, Ramola Rakesh, Masahide Furukawa: A comparative study of thorium activity in NORM and high background radiation area, Radiation Protection Dosimetry (2010).
10. Michikuni Shimo, Yuu Ishimori, Masahiro Hosoda, Tokonami, S.: Thoron exhalation rate in many areas in Japan, Radiation Protection Dosimetry (2010).
11. Yoshitaka Shiroma, Noki Isa, Masahiro Hosoda, Atsuyuki Sorimachi, Tetsuo Ishikawa, Tokonami, S., Masahide Furukawa: In situ measurement on thoron exhalation rate in Okinawa, Japan, Radiation Protection Dosimetry (2010).
12. Atsuyuki Sorimachi, Tetsuo Ishikawa, Janik Miroslaw, Tokonami, S.: Quality assurance and quality control for thoron measurement at NIRS, Radiation Protection Dosimetry (2010).

13. Janik Miroslaw, Tokonami, S., Kranrod Chutima, Atsuyuki Sorimachi, Tetsuo Ishikawa, Mohamed Hassan Nabil Mohamed: International intercomparisons of integrating radon/thoron detectors with the NIRS radon/thoron chambers, Radiation Protection Dosimetry (2010).
14. Tokonami, S.: Why is 220Rn (thoron) measurement important?, Radiation Protection Dosimetry (2010).
15. C. Németh, V. Jobbágy, N. Kávási, J. Somlai, T. Kovács and S. Tokonami: Radon and Thoron parallel measurements in dwellings nearby a closed Hungarian uranium mine, Nukleonica (2010)
16. N. M. Hassan, M. Fukushi, S. Tokonami: A simple technique for studying the dependence of radon and thoron exhalation rate from building materials on absolute humidity, Journal of Radioanalytical and Nuclear Chemistry, doi: 10.1007/s10967-010-0665-7 (2010)
17. S. K. Sahoo, M. Hosoda, A. Sorimachi, T. Ishikawa, S. Tokonami, S. Uchida: Thorium, uranium and rare earth elements concentration in weathered Japanese soil samples, Progress in Nuclear Science and Technology (2010)
18. N. M. Hassan, M. Hosoda, K. Iwaoka, A. Sorimachi, J. Miroslaw, K. Chutima, S. K. Sahoo, T. Ishikawa, H. Yonehara, M. Fukushi, S. Tokonami: Simultaneous Measurement of Radon and Thoron Released from Building Materials Used in Japan, Progress in Nuclear Science and Technology (2010)
19. T. Ishikawa, M. Hosoda, A. Sorimachi, S. Tokonami, S. Katoh, S. Ogashiwa: Radiological characterization of commercially available radon spa sources, Journal of Radioanalytical and Nuclear Chemistry (2010)
20. N. M. Hassan, M. Fukushi, Tokonami, S.: A simple technique for studying the dependence of radon and thoron exhalation rate from building materials on absolute humidity, Journal of Radioanalytical and Nuclear Chemistry, doi: 10.1007/s10967-010-0665-7 (2010)
21. J. Chen, B. Walker, A. Sorimachi, H. Takahashi, S. Tokonami: An investigation on radon and thoron response of alpha-track detectors used in the Winnipeg case-control study, Radiation Protection Dosimetry, 138(1), 83-86, 2010
22. N. M. Hassan, T. Ishikawa, M. Hosoda, A. Sorimachi, S. Tokonami, M. Fukushi, Sahoo Sarata Kumar: Assessment of the natural radioactivity using two techniques for the measurement of radionuclide concentration in building materials used in Japan, Journal of Radioanalytical and Nuclear Chemistry, 283, 15-21, 2010
23. Y. Yasuoka, T. Ishikawa, N. Fukuhori, S. Tokonami: Comparison of simplified liquid scintillation counter (Triathler) with conventional liquid scintillation counter in the measurement of radon concentration in water, 59(1), 11-21, 2009
24. A. Sorimachi, S. K. Sahoo, S. Tokonami: Generation and control of thoron emanated from lantern mantles, Review of Scientific Instruments, 80(1), 015104-1-015104-4, 2009
25. M. Hosoda, S. Tokonami, A. Sorimachi, J. Miroslaw, T. Ishikawa, Y. Yatabe, J. Yamada, S. Uchida: Experimental system to evaluate the effective diffusion coefficient of radon, Review of Scientific Instruments, 80(1), 013501, 2009
26. Z. S. Zunic, Y. Ilia, V. Nenad, Z. Michael, P. Ujic, I. Celikovic, J. McLaughlin, S. Simos, B. Aleksandar, K. Fujimoto, P. Johan, T. Flavio, S. Tokonami, O. Pawel, K. Kryzstof, B. Francesco, R. Rakesh, M. Jerzy, J. Bajram, M. Gordana, C. Giancarlo, K.

Kevin, B. Macej, S. K. Sahoo, V. Hans, W. Micael: Identification and assessment of elevated exposure to natural radiation in Balkan region (Serbia), Radioprotection, 44(5), 919-925, 2009

27. S. K. Sahoo, S. Yoshida, S. Tokonami, H. Yonehara et.al: Isotopic composition of Uranium in soil and ground water samples collected around 30 km Chernobyl Exclusion Zone, Radioprotection, 44(5), 785-790, 2009
28. K. Chutima, C. Supitcha, C. Nares, S. Tokonami, T. Ishikawa, S. K. Sahoo: A simple technique for determining the equilibrium equivalent Thoron concentration using a CR-39 detecter: Application in mineral treatment industry, Radioprotection, 44(5), 301-304, 2009
29. A. Sorimachi, K. Chutima, S. Tokonami, T. Ishikawa, M. Hosoda, J. Miroslaw, R. Shingaki, M. Furukawa: Anomalously High Radon Concentrations in a Dwelling in Okinawa, Japan, Radioisotopes, 58(12), 807-813, 2009
30. J. Chen, E. Schroth, E. Mackinlay, I. Fife, A. Sorimachi, S. Tokonami: Simultaneous ^{222}Rn and ^{220}Rn measurements in Winnpeg,Canada., Radiation Protection Dosimetry, 134(2), 75-78, 2009
31. B. Francesco, A. Ampollini, L. Tommasino, A. Sorimachi, S. Tokonami: Sensitivity to thoron of an SSNTD-based passive radon measuring device: Experimental evaluation and implications for radon concentration measurements and risk assessment, Radiation Measurements, 44(9-10), 1024-1027, 2009
32. N. Kavasi, J. Somlai, V. Tamas, S. Tokonami, T. Ishikawa, A. Sorimachi, T. Kovacs: Difficulties in the dose estimate of workers originated from radon and radon progency in a manganese mine., Radiation Measurements, 44(3), 300-305, 2009
33. A. Sorimachi, H. Takahashi, S. Tokonami: Influence of the presence of humidity, ambient aerosols and thoron on the detection responses of electret radon monitors., Radiation Measurements, 44(1), 111-115, 2009
34. Y. Yasuoka, Y. Kawada, H. Nagahama, Y. Omori, T. Ishikawa, S. Tokonami, M. Shinogi: Preseismic changes in atmospheric radon concentration and crustal strain, Physics and Chemistry of the Earth, 34(6/7), 431-434, 2009
35. Y. Omori, H. Nagahama, Y. Kawada, Y. Yasuoka, T. Ishikawa, S. Tokonami, M. Shinogi: Preseismic alteration of atmospheric electrical conditions due to anomalous radon emanation, Physics and Chemistry of the Earth, 34(6-7), 435-440, 2009
36. Y. Yasuoka, T. Ishikawa, S. Tokonami, H. Takahashi, A. Sorimachi, M. Shinogi: Radon mitigation using an air cleaner, Journal of Radioanalytical and Nuclear Chemistry, 279(3), 885-891, 2009
37. T. Ishikawa, Y. Yasuoka, S. Tokonami, H. Takahashi, H. Suda: Characterization of airborne radon released by domestic showering and the use of a washing machine, Journal of Radioanalytical and Nuclear Chemistry, 279(3), 719-724, 2009
38. J. Somlai, S. Gabor, P. Szabo, A. Varhegyi, S. Tokonami, T. Ishikawa, A. Sorimachi, S. Yoshinaga, T. Kovacs: Radiation dose of workers originating from radon in the show cave of Tapolca, Hungary, Journal of Radioanalytical and Nuclear Chemistry, 279(1), 219-225, 2009
39. K. Doi, S. Tokonami, H. Yonehara, S. Yoshinaga: A Simulation Study of Radon and Thoron Discrimination Problem in Case-Control Studies, Journal of Radiation Research, 50(6), 495-506, 2009

40. Z. S. Zunic, J. Miroslaw, S. Tokonami, V. Nenad, Y. Ilia, Z. Michael, T. Ishikawa, R. Rakesh, C. Giancarlo, J. Peter, K. Kryzstof, M. Jadwiga, I. Celikovic, P. Ujic, O. Aleksandra, S. K. Sahoo, B. Francesco: Field Experience with Soil Gas Mapping Using Japanese Passive Radon/Thoron Discriminative Detectors for Comparing High and Low Radiation Areas in Serbia (Balkan Region), Journal of Radiation Research, 50(4), 355-361, 2009
41. M. Hosoda, A. Sorimachi, Y. Yasuoka, T. Ishikawa, S. K. Sahoo, M. Furukawa, N. M. Hassan, S. Tokonami, S. Uchida: Simultaneous Measurements of Radon and Thoron Exhalation Rates and Comparison with Values Calculated by UNSCEAR Equation, Journal of Radiation Research, 50(4), 333-343, 2009
42. J. Miroslaw, S. Tokonami: Natural and Artificial Sources of Radioactivity in Poland, Japanese Journal of Health Physics, 44(1), 116-121, 2009
43. N. M. Hassan, M. Hosoda, T. Ishikawa, S. Tokonami, M. Fukushi et.al: ^{222}Rn exhalation rate from Egyptian building materials using active and passive methods, Japanese Journal of Health Physics, 44(1), 106-111, 2009
44. S. Tokonami: Thoron in the environment and its related issues, Indian Journal of Physics, 83(6), 777-785, 2009
45. J. Miroslaw, S. Tokonami, T. Kovacs, N. Kavasi, K. Chutima, A. Sorimachi, H. Takahashi, T. Ishikawa: International intercomparisons of integrating radon detectors in the NIRS radon chamber., Applied Radiation and Isotopes, 67(9), 1691-1696, 2009
46. K. Chutima, S. Tokonami, T. Ishikawa, A. Sorimachi, J. Miroslaw, R. Shingaki, M. Furukawa, C. Supitcha, C. Nares: Mitigation of the effective dose of radon decay products through the use of an air cleaner in a dwelling in Okinawa, Japan., Applied Radiation and Isotopes, 67(6), 1127-1132, 2009
47. S. Yoshinaga, T. Ishikawa, S. Tokonami et.al: Radon in drinking water and cancer mortality: an ecological study in Japan, The Natural Radiation Environment : 8th International Symposium (NRE VIII), Buzios, Rio de Janeiro, Brazil, 7-12 October 2007 (AIP Conference Proceedings ; vol. 1034), 1034, 429-432, 2008
48. T. Ishikawa, S. Tokonami, Y. Kobayashi, A. Sorimachi, Y. Yatabe, N. Miyahara: Evaluation of gas-filled ionization chamber method for radon measurement at two reference facilities, The Natural Radiation Environment : 8th International Symposium (NRE VIII), Buzios, Rio de Janeiro, Brazil, 7-12 October 2007 (AIP Conference Proceedings ; vol. 1034), 1034, 423-426, 2008
49. S. K. Sahoo, S. Tokonami, T. Ishikawa et.al: Determination of depleted uranium in environmental bio-monitor samples and soil from target sites in western Balkan region, The Natural Radiation Environment : 8th International Symposium (NRE VIII), Buzios, Rio de Janeiro, Brazil, 7-12 October 2007 (AIP Conference Proceedings ; vol. 1034), 1034, 287-290, 2008
50. A. Sorimachi, S. Tokonami, H. Takahashi, Y. Kobayashi: Performance of NIRS thoron chamber system, The Natural Radiation Environment : 8th International Symposium (NRE VIII), Buzios, Rio de Janeiro, Brazil, 7-12 October 2007 (AIP Conference Proceedings ; vol. 1034), 1034, 206-209, 2008
51. S. Tokonami, T. Ishikawa, A. Sorimachi, H. Takahashi, N. Miyahara: The Japanese radon and thoron reference chambers, The Natural Radiation Environment : 8th International Symposium (NRE VIII), Buzios, Rio de Janeiro, Brazil, 7-12 October 2007 (AIP Conference Proceedings ; vol. 1034), 1034, 202-205, 2008

52. M. Oka, M. Shimo, S. Tokonami, A. Sorimachi, H. Takahashi, T. Ishikawa: Measurement of indoor radon-222 and radon-220 concentrations in central Japan, The Natural Radiation Environment : 8th International Symposium (NRE VIII), Buzios, Rio de Janeiro, Brazil, 7-12 October 2007 (AIP Conference Proceedings ; vol. 1034), 1034, 189-192, 2008
53. T. Ishikawa, S. Tokonami, Q. Sun, Y. Kobayashi, S. Yoshinaga et.al: Preliminary results of indoor radon/thoron concentrations and terrestrial gamma doses in Gejiu, Yunnan, China, The Natural Radiation Environment : 8th International Symposium (NRE VIII), Buzios, Rio de Janeiro, Brazil, 7-12 October 2007 (AIP Conference Proceedings ; vol. 1034), 1034, 173-176, 2008
54. O. N. Choncubhair, J. McLaughlin, S. Tokonami: Passive measurements of thoron and its progeny in some dwellings in Ireland, The Natural Radiation Environment : 8th International Symposium (NRE VIII), Buzios, Rio de Janeiro, Brazil, 7-12 October 2007 (AIP Conference Proceedings ; vol. 1034), 1034, 16-19, 2008
55. S. Tokonami, S. Yoshinaga, M. Sugino, Y. Kobayashi, H. Takahashi, A. Sorimachi, T. Ishikawa et.al: Influence of environmental thoron on radon measurements and related issues, The Natural Radiation Environment : 8th International Symposium (NRE VIII), Buzios, Rio de Janeiro, Brazil, 7-12 October 2007 (AIP Conference Proceedings ; vol. 1034), 1034, 145-148, 2008
56. J. Chen, S. Tokonami, A. Sorimachi, H. Takahashi, R. Falcomer: Preliminary results of simultaneous radon and thoron tests in Ottawa, Radiation Protection Dosimetry, 130(2), 253-256, 2008
57. Y. Yasuoka, T. Ishikawa, S. Tokonami, H. Takahashi, Y. Narazaki et.al: A case study on the effect of water from groundwater sources on indoor radon levels, Journal of Radioanalytical and Nuclear Chemistry, 275(1), 165-172, 2008
58. S. Tokonami, K. Tibor, S. Yoshinaga, Y. Kobayashi, T. Ishikawa: ^{210}Po and ^{210}Pb inhalation dose by cigarette smoking in Gansu and Yunnan provinces, China, Japanese Journal of Health Physics, 43(2), 131-134, 2008

ISBN: 978-1-62100-771-5

Chapter 84

BIOGRAPHICAL SKETCH

Name:	Title:	Date of Birth:
Vukosava Milic Torres	National Institute of Health Dr Ricardo Jorge, Department of Genetics, Laboratory of proteomics, Av. Padre Cruz 1649-016 Lisbon, Portugal	October 13, 1968

EDUCATION:

Institution and Location	Degree	Year Conferred	Scientific Field
Pharmacy; subdicipline-pharmacology/toxicology Title: "Examination of cardioprotective efficacy and protective mechanisms of fullerenol C60(OH)24 in doxorubicin treated rats",Military Medical Academy, Department of Toxicology, Belgrade, Serbia	PhD	2007	
Analytical chemistry, "Simultaneous extraction of organochlorine pesticides from soil using SPE and it determination by Gas Chromatography-Mass Spectrometry", Department of Chemistry, Faculty of sciences, University of Novi Sad, Serbia	MSci thesis	2001	

Institution and Location	Degree	Year Conferred	Scientific Field
"Voltametric characterization of some new Fe(III) chelates in non aqueous media", Department of Chemistry, Faculty of Sciences, University of Novi Sad,Serbia	Licenced in Chemistry	1994	

CONTACT POINTS:

Av. Padre Cruz, 1649-016 Lisbon, Portugal

RESEARCH AND PROFESSIONAL EXPERIENCE:

2010- Translational proteomics and biomarker research 2009- Potentially pharmacologically active fullerene derivatives based on C60-phenolic conjugates: Solubilization and investigation of metabolism, permeability, toxicity and biological activity

2003 – 2008 Teaching medicinal chemistry and toxicology; responsible investigator for method development in clinical trials (analytical method development, validation); responsible investigator for biological examination of fullerenol on the project " Synthesis of bioactive fullerene molecules and nanomedical research" Grant No 142076, Ministry of Sciences, Republic of Serbia

2002 – 2003 Managing laboratory services in the field of quality control of food, water, waste water, animal food and environmental quality. Responsible for developing, implementing and validation of new analytical methods. Responsible for system of quality ISO 9001:2000 and ISO/IEC 17025.

1997 – 2002 Chemical analyst in Institute of Public Health, Novi Sad, Department of Hygiene, Laboratory for quality control of food, cosmetics, water and waste water. Working on determination of heavy metals, pesticides, PCBs and PAH in different samples. Analysis of pesticides, preservatives, mycotoxines, heavy metals in food, cosmetics, water, waste water, environment (GC/MS; SPE, GC/ECD/FID, SFE, UV/VIS, HPLC/DAD/FLD)

1994 – 1997 Working as a junior researcher in soil chemistry, determination of heavy metals, pesticides and PCBs in soil.

Professional Appointments:

2010- Postdoctoral fellow, National Institute of Health Dr Ricardo Jorge, Department of Genetics, Laboratory of proteomics, Portugal

2009-2010 Postdoctoral fellow, ITQB, Laboratory for Pharmacokinetics and Pharmacutical analysis, Oeiras, Portugal 2008- Assistant professor of Toxicology Department of Pharmacy, Medical faculty Novi Sad, University of Novi Sad, Serbia (position on hold)

2003 – 2008 Teaching assistant on Toxicology and Pharmaceutical analysis, Department of Pharmacy, Medical faculty, University of Novi Sad, Serbia

2002 – 2003 Head of ecotoxicological laboratory of "Bioecological center" Ltd. Zrenjanin, Serbia

1997 – 2002 Institute of Public Health, Novi Sad, Department of Hygiene, Laboratory for quality control of food, cosmetics, water and waste water. Novi Sad, Serbia

1994 – 1997 Institute of field and vegetable crops in Novi Sad, Serbia

Publications during last three years:

- Milic Torres V, Posa M, Srdjenovic B, Simplício AL. Solubilization of fullerene C60 in micellar solutions with different solubilizers. Colloids and Surfaces B: Biointerfaces. 2011; 82:46-53.
- Srdjenovic B. , Milic Torres V, Grujic N, Stankov K, Djordjevic A, Vasovic V. Antioxidant properties of fullerenol $C_{60}(OH)_{24}$ in rat kidneys, testes and lungs treated with doxorubicin. Tox. Mech. Meth. 2010; 20(6): 298-305.
- Djordjevic Milic V,Srdjenovic B , Jacevic V, Dragojevic Simic V, Djordjevic A and Simplício AL.Fullerenol $C_{60}(OH)_{24}$ successfully prevents doxorubicin induced acute cardiotoxicity in rats.Pharmacological Reports, 2010, vol. 62, no. 707 – 718.
- Djordjevic Milic V, Stankov K, Injac R, Djordjevic A, Srdjenovic B, Govedarica B, Radic N, Dragojevic Simic V, Strukelj B. Activity of Antioxidative Enzymes in Erythrocytes after a Single Dose Administration of Doxorubicin in Rats Pretreated with Fullerenol C60(OH)24. Toxicology mechanisms and methods 2009;19: 24-28.
- Injac R, Perse M, Obermajer N, Djordjevic Milic V, Prijatelj M, Djordjevic A, Cerar A, Strukelj B. Potential hepatoprotective effects of fullerenol C60(OH)24 in doxorubicin-induced hepatotoxicity in rats with mammary carcinomas. Biomaterials 2008; 29: 3451-3460.
- Injac R, Strukelj B, Djordjevic Milic V, Karljikovic Rajic K, Lingeman H. Densitometric determination of zinc bacitracin and nystatin in animal feed. Journal of the Science of Food and Agriculture 2008; 88: 1576-1581.
- Injac R, Perse M, Boskovic M, Djordjevic-Milic V, Djordjevic A, Hvala A, Cerar A, Strukelj B. Cardioprotective Effects of Fullerenol $C_{60}(OH)_{24}$ on a Single Dose Doxorubicin-induced Cardiotoxicity in Rats With Malignant Neoplasm. Technology in Cancer Research & Treatment 2008; 7: 15-26
- Vasović V, Banić B, Jakovljević V, Djordjevic Milic V. Effect of aminophyline on aspirin transition into central nervous system in rats. European Journal of Drug Metabolism and Pharmacokinetics. 2008; 33(1): 23-30.
- Srdjenovic B, Djordjević Milić V, Grujic N, Injac R, Lepojevic Z. Simultanious HPLC determination of caffeine, theobromine and theophylline in food, drinks and herbal products. Journal of Chromatographic Science 2008; 46: 144-149
- Injac R, Djordjević Milić V, Mlinaric A, Strukelj B. Optimal condition for determination of zink bacitracin, polymyxin B, oxytetracycline and sulfacetamide in animal feed by micelar electrokinetic capillary chromatography. Food aditives and contaminants 2008; 25(4): 424-431

In: Chemistry Researcher Biographical Sketches
Editors: E. S. Huseynov and V. Babayev
ISBN: 978-1-62100-771-5

Chapter 85

BIOGRAPHICAL SKETCH

Name:	Title:	Date of Birth:
Rolando J. Tremont	University of Puerto Rico, Humacao Campus	July 20, 1964

EDUCATION:

Institution and Location	Degree	Year Conferred	Scientific Field
Analytical Chemistry	Ph.D		

CONTACT POINTS:

Department of Chemistry, University of Puerto Rico, CUH Station, 100 Road 908, Humacao 00791-4300.

RESEARCH AND PROFESSIONAL EXPERIENCE:

Studies electrochemical and Surface Analysis in the modification of metal surface with organic compounds, using seft-assembled monolayer (SAM) technique. These surfaces can be used in catalysis, corrosion inhibitors, control of metallic nanoparticles, biosensors and growth of crystals.

1992-1993 Chemist, Planta de Fraccionamiento del Petróleo Jose, CORPOVEN. Anzoátegui, Venezuela
1993-1997 Lecturer, Chemistry Department, Universidad de Oriente, Venezuela
2002-2005 Assistant Professor, Chemistry Department, University of Puerto Rico, Río Piedras

Summer 2005 Invited Professor, Chemistry Dept., Clemson University, SC
2005- Present Associate Professor, Chemistry Department, University of Puerto Rico, Humacao
2007- Present Director, Chemistry Department, University of Puerto Rico, Humacao.

Professional Appointments:

2007-Present: American Association for the Advancement of Science (AAAS).
1997-Present: Electrochemistry International Society
1997-Present: American Chemical Society (ACS) (Member No: 2214317)
1997-Present: The Electrochemical Society, Inc. (Member No:: 110119)
1998-Present: National Associate of Corrosion (NACE International) (Member No: 154825-00)
2000-2002 NSF-EPSCoR (National Science Foundation-Experimental Program to Stimulate Competitive Research).
2004 Prize to the best research Work presented at "XII Encuentro de Electroquímica", Venezuela. Title: "Systematic Study of Corrosion Damage caused by Petroleum Separation".

Publications during last three years:

Tremont, R.J., Cabrera, C.R, De Jesús-Cardona H., and Castro, R.J. 2000. (3-Mercaptopropyltrimethoxysilane as Cu Corrosion Inhibitor in KCl Solution. J. Applied Electrochemistry. 30:737-749.

Cabrera, C.R, Warner J.D, Mueller C.H, Van Keuls F., Miranda F.A, Tremont R.J, Blasini, D.R, and Morales A. 2001. Ferroelectric Charge Injection MOSFET Devices. Integrated Ferroelectrics. 38: 269.

Cabrera, C.R., Warner, J.D., Mueller. C.H., Van Keuls, F., Miranda, F.A., Tremont, R.J., Blasini, D., and Morales, A.. 2001. Self-Assembled 3-Mercaptopropyltrimethoxysilane (MPS) on Ba0.5Sr0.5TiO3 as an Adhesion Layer for Microwave Devices. Materials Research Society (MRS) Proceedings.666, F9.1/1-F9.1/6.

Tremont R.J., and Cabrera C.R. 2002. Corrosion Protection Comparison of Copper by a Self assembled Monolayer of 1-Propanethiol and Propyltrimethoxysilane: A comparison with 3 mercaptopropyltrimethoxysilane. J. of Applied Electrochemistry. 32: 783-795.

Blasini D.R, Tremont R.J., and Cabrera C.R, Batina, N., and González, I. 2003. Self-Assembled (3-Mercaptopropyl)trimethoxysilane on Iodine Coated Gold Electrodes. J. Electroanalytical Chemistry. 540: 45-52.

Brito, R, Tremont R.J., Feliciano, O., and Cabrera, C.R. 2003. Chemical Derivatization of Self-Assembled 3-Mercaptopropionic and 16-Mercaptodecanoic Acids at Platinum Surface with 3-Aminopropyltrimethoxysilane: A Spectroscopic and Electrochemical Studies. J. Electroanalytical Chemistry. 540: 53-62.

Tremont, R.J., Blasini, D., and Cabrera, C.R. 2003. Controlled Self-Assembly of Mercapto and Silane Terminated Molecules at Cu Surfaces. Journal Electroanalytical Chemistry. 556:147-158.

Tremont, R.J., Cruz, G., and Cabrera, C.R. 2003. Pt nanoparticles on Cu surface modified with 3-mercaptopropyltrimethoxysilane and 1-propanethiol. J. Electroanalytical Chemistry. 558: 65-77.

Bolívar, H., Izquierdo, S., Tremont, R.J., and Cabrera, C.R. 2003 Pt/MoOx/MoSe2 Electrodes Prepared from Li Intercalated-Exfoliated MoSe2: Applications in Methanol Oxidation. J. Appl. Electrochemistry. 33: 1191-1198.

Sosa, E., Cabrera-Sierra, R., Oropeza, M.T, Hernández, F., Casillas, N., Tremont, R.J. Cabrera, C.R, and González, I. 2003 Chemical Characterization of Corrosion Films Electrochemically Grown on Carbon Steel in Alkaline Sour Environment. J. of The Electrochemical Society. 150 (11), B530-B535.

Sosa, E., Cabrera-Sierra, R., Oropeza, M.T., Hernández, F., Casillas, N., Tremont R.J., Cabrera, C.R, and Gonzalez, I. 2003 Electrochemically grown passive films on carbon steel (SAE 1018) in alkaline sour medium. Electrochimica Acta. 48 (12), 1665-1676.

Brito, R., Tremont, R.J., and Cabrera, C.R. 2004. Study of the electron transfer kinetics across a derivatized self assembled monolayer on platinum by cyclic voltammetry and electrochemical impedance spectroscopy. J. Electroanalytical Chemistry. 574: 15-22.

Morales-Cruz, A.,L, Tremont, R.J., Martínez, R., Romañach, R., and Cabrera, C.R.. 2005. Atomic force measurements of 16-mercaptohexadecanoic acid and its salt with CH3, OH, and CONHCH3 functionalized self-assembled monolayers. Applied Surface Science. 241(3-4): 371-383.

Markus Frei , François Diederich, Rolando Tremont, Tanya Díaz , Luis Echegoyen, 2006, Tetrathiafulvalene- (TTF) bridged Resorcin[4]arene Cavitands: Towards New Electrochemical Molecular Switches, Helvetica Chimica Acta, 89 (9) 2040 – 2057.

Tanya L. Díaz, María Malavé, Mirna Rivera, Jorge Castillo, Carlos R. Cabrera, Rosa Brito, Rolando J. Tremont, 2008, Modification of Au Surfaces Using New Ferrocene Derivatives, Applied Surface Science. 254 (6), 1587-1592.

Luisa Rojas de Astudillo, Linoshka Rivera, Rosa Brito-Gómez, and Rolando J. Tremont, 2010, Electrochemical study of 1,4-Benzoquinone on Gold Surface Modified, J. Electroanallytical Chemistry, 640, 56-60.

In: Chemistry Researcher Biographical Sketches
Editors: E. S. Huseynov and V. Babayev
ISBN: 978-1-62100-771-5

Chapter 86

BIOGRAPHICAL SKETCH

Name:	Title:	Date of Birth:
Hirotada Tsujii		January 13, 1945

EDUCATION:

Institution and Location	Degree	Year Conferred	Scientific Field
Tohoku University graduate school doctor's course			

CONTACT POINTS:

9636-5 Minamiminowa-mura, Kamiina-gunn, Nagano-ken, 399-4511 JAPAN

RESEARCH AND PROFESSIONAL EXPERIENCE:

38 year, Professor, Shinshu University:15year

Professional Appointments:

Professor emeritus, Shinshu University

Awards and Honors:

Professor emeritus, Shinshu University

Publications during last three years:

S. Afrose, M.S. Hossain, T.Maki and H.Tsujii,: Karaya root saponin exerts a hypocholesterolemic response in rats fed a high-cholesterol diet. Nutrition Research 2009,29:250-354.

S. Afrose, M.S. Hossain and H.Tsujii,: Effects of dietary karaya saponin and Rhodobacter capsulatus on yolk cholesterol in laying hens. British Poultry Science 2010,51:409-418

S. Afrose, M.S. Hossain and H.Tsujii,:Effect of dietary karaya saponin on serum and egg yolk cholesterol in laying hens. British Poultry Science 2010,51:797-804

S. Afrose, M.S. Hossain and H.Tsujii,: Hypocholesterolemic effect of karaya saponin in Japanese laying quails (Coturnic coturnix japonica). J Animal Physiology and Animal Nutrition 2010 Nov 28. doi: 10.1111/j.1439-0396.2010.01098.x.

Afrose S, Hossain MS, Salma U, Miah AG, Tsujii H. Dietary Karaya Saponin and Rhodobacter capsulatus Exert Hypocholesterolemic Effects by Suppression of Hepatic Cholesterol Synthesis and Promotion of Bile Acid Synthesis in Laying Hens. Cholesterol. 2010;2010:272731.

Md. Sharoare Hossain, Sadia Afrose, Iwao Takeda1 and Hirotada Tsujii Effect of Selenium-enriched Japanese Radish Sprouts and *Rhodobacter capsulatus* on the Cholesterol and Immune Response of Laying Hens. Asian-Aust. J. Anim. Sci. Vol. 23, No. 5 : 630 - 639 2010

Hossain MS, Afrose S, Sawada T, Hammano K, Tsujii H. Metabolism of exogenous fatty acids, fatty acid- mediated cholesterol efflux, PKA and PKC pathways in boar sperm acrosome reaction. Reproductive Medicine and Biology, 2010; 9:23-31.

R. Obata and H. Tsujii: Effects Of Amino Acids On The In Vitro Development Of Mongolian Gerbil (meriones Unguiculatus) Embryos .J.fMammalian Ova Res. 27(3):122-129. 2010

R. Obata and H. Tsujii: Valine Metabolism Of Mongolian Gerbil (meriones Unguiculatus) Embryos J. Mam. Ova. Res. 27(3): 144-149. 2010

Miah AG, Salma U, Sinha PB, Hölker M, Tesfaye D, Cinar MU, Tsujii H, Schellander K. Intracellular signaling cascades Anim Reprod Sci. 2011 Mar 23. [Epub ahead of print]

H.Tsujii Effects of glucose and fructose on development of mouse and rat embryos in vitro. Nova Science Publishers, Inc; ISBN:978-1-61761-263-3 2011

In: Chemistry Researcher Biographical Sketches
Editors: E. S. Huseynov and V. Babayev
ISBN: 978-1-62100-771-5

Chapter 87

BIOGRAPHICAL SKETCH

Name:	Title:	Date of Birth:
Tiverios C. Vaimakis	Department of Chemistry University of Ioannina 45110 Ioannina, Greece	31 May 1953

EDUCATION:

Institution and Location	Degree	Year Conferred	Scientific Field
PhD in Chemistry			

CONTACT POINTS:

Department of Chemistry, University of Ioannina, 45110 Ioannina, Greece

RESEARCH AND PROFESSIONAL EXPERIENCE:

Thin films and Powders, Bioceramics, Thermal Analysis, Beneficiation of phosphate rocks, Mechanism of heterogeneous processes

Professional Appointments:

Associate Professor

Publications during last three years:

1. T.C. Vaimakis, E.D. Economou, and C.C. Trapalis, Calorimetric Study of Dissolution Kinetics of Phosphorite in Diluted Acetic Acid• , *J. Thermal Anal. Calorim.*, 92, 783 (2008).
2. George I. Chilas, Marios Stylianou, Maciej Kubicki, Tiverios Vaimakis, Paul Kögerler, Anastasios D. Keramidas, and Themistoklis A. Kabanos, "Polynuclear Cobalt(II/III) Sulfites: Synthesis, Structure, and Magnetic Properties of the Octanuclear Cluster $(NH_4)_{11}$(Li ℘□ $[Co_4^{II}Co_4^{III}(SO_3\ )_{16}(NH_3)_8])\cdot 10H_2O$ Encapsulating a Lithium Cation• , *Inorg. Chem. (Communication)*,47, 4451 (2008)
3. Sindou E., Vaimakis V., Vaimakis T., Roussis I.G., Effect of juice clarification by flotation on the quality of white wine and orange juice and drink - Short communication• , *Czech J. Food Sci.*, 26 (3): 223-228 2008
4. N. Todorova, T. Giannakopoulou, T. Vaimakis, G. Romanos and J. Yu, C. Trapalis, "Preparation of fluorine doped TiO2 photocatalysts with controlled crystalline structure" *Inter. J. Photoenergy*, Art. No. 534038 2008
5. T. Giannakopoulou, N. Todorova, T. Vaimakis, S. Ladas, C. Trapalis, Study of fluorine doped TiO_2 sol-gel thin coatings• *Journal of Solar Energy and Engineerin , 130 (4), Art. 041007, 2008*
6. A. Giannakas, C.G. Spanos, N. Kourkoumelis, T. Vaimakis, A. Ladavos, Preparation, characterization and water barrier properties of PS/organo-montmorillonite nanocomposites• , European Polymer Journal, 44 (12), 3915-3921, 2008
7. N. Todorova, T. Giannakopoulou, T. Vaimakis, C. Trapalis, Structure Tailoring of Fluorine-doped TiO_2 Nanostructured Powders• *Materials Science and Engineering B-Advanced Functional Solid-State Materials* 152 (1-3), pp. 50-54, 2008
8. A. Giannakas, C. G. Spanos, N. Kourkoumelis, T. Vaimakis and A. Ladavos, Structure and Thermal Stability of Polystyrene/Layered Silicate Nanocomposites• , Composite Interfaces 16, pp 237–247, 2009
9. T. Giannakopoulou, N. Todorova, P. Osiceanu, A. Lagoyannis, T. Vaimakis and C. Trapalis, Description of TiO_2 thin films treated in NH_3 atmosphere by optical dispersion models• , Thin Solid Films 517 (2009) 6694-6699
10. Anastasios I. Mitsionis, Tiverios C. Vaimakis, A Calorimetric Study of the temperature effect on Calcium Phosphate precipitation• , *J. Thermal Anal. Calorim,* 99 (2010),785–789.
11. A.J. Mitsionis, T.C. Vaimakis, and C.C. Trapalis, « The effect of citric acid on the sintering of the calcium phosphate bioceramics• , *Ceramic International*, 36 (2010), 623-634
12. G.I. Chilas, N. Lalioti, T. Vaimakis, M. Kubicki, and Th. Kabanos, Hydrothermal syntheses, crystal structures and physicochemical properties of 2-D and 3-D inorganic coordination cobalt(II)-sulfite polymers• , *Dalton Trans.*, 39 (35)2010, pp. 8296-8305
13. K. Mahmud, T.C. Vaimakis, A.J. Mitsionis, and C.C. Trapalis, The Threonine Effect on Calcium Phosphate Preparation from a Solution Containing Ca/P=1.33 Molar Ratio• , *Ceramic International*, 36 (2010) 1893–1899

14. D. Karamanis, A. N. Ő¦kte, E. Vardoulakis, T. Vaimakis, Photo-responsive TiO2-sepiolite nanocomposite for building integrated evaporative cooling and pollutant degradation• , Applied Clay Science, (in press).

In: Chemistry Researcher Biographical Sketches
Editors: E. S. Huseynov and V. Babayev
ISBN: 978-1-62100-771-5

Chapter 88

BIOGRAPHICAL SKETCH

Name:	Title:	Date of Birth:
Bakharev Vladimir Valentinovich	Samara State Technical University	02-03-1965

EDUCATION:

Institution and Location	Degree	Year Conferred	Scientific Field
Chemical Science (PhD) at Samara State Technical University	doctor's degree	2008	
Chemical Science at Samara State Technical University	candidate's degree	1994:	
Chemical Technology of Nitrogen Organic Compounds at Kuibyshev Polytechnic Institute	five-year University degree	1990	

CONTACT POINTS:

244, Molodogvardeyskaya st., Samara, 443100, Russia

RESEARCH AND PROFESSIONAL EXPERIENCE:

Senior research scientist at research laboratory of Samara State Technical University

Professional Appointments:

Professor of the Chemistry and Technology of Nitrogen Organic Compounds chair

Publications during last three years

1. Gidaspov, V. V. Bakharev and I. K. Kukushkin, Trinitromethylation reactions of chloro-1,3,5-triazines, Russian Chemical Bulletin, 2009, Vol. 58, Iss.10, 2154-2163.
2. V. V. Bakharev, A. A. Gidaspov, E. V. Peresedova, V. G. Granik and N. B. Grigor'ev, et al., 1,3,5-Triazinenitrolic acids. Synthesis and NO-releasing properties, Russian Chemical Bulletin, 2009, Vol. 58, Iss.9, 1962-1972.
3. Gidaspov, V. V. Bakharev, B. S. Fedorov, M. A. Fadeev and N. P. Konovalova, New antimetastatic preparations based on chlorodinitromethyl-1,3,5-triazines, Russian Journal of Applied Chemistry, 2009, Vol. 82, Iss. 10, 1816-1820.
4. V. V. Bakharev, A. A. Gidaspov, E. V. Selezneva, I. V. Ul'yankina, D. B. Krivolapov, I. A. Litvinov and O. S. Eltsov, Reactions of 1,3,5-triazinylnitroformaldoximes 3*. Interaction of 1,3,5-triazinylnitroformaldoximes with malonic acid esters, Chemistry of Heterocyclic Compounds, Vol. 46, Iss. 11, 1386-1392

In: Chemistry Researcher Biographical Sketches
Editors: E. S. Huseynov and V. Babayev
ISBN: 978-1-62100-771-5

Chapter 89

BIOGRAPHICAL SKETCH

Name:	Title:	Date of Birth:
Konarev Dmitri Valentinovich	Institute of Problems of Chemical Physics RAS	24.02.1970

EDUCATION:

Institution and Location	Degree	Year Conferred	Scientific Field
Diploma of Chemistry St. Petersburg State University		1993	
Institute of Problems of Chemical Physics RAS	Ph.D.,	1999	

CONTACT POINTS:

Institute of Problems of Chemical Physics Russian Academy of Science, Chernogolovka, Moscow region, 142432, Russia

RESEARCH AND PROFESSIONAL EXPERIENCE:

Physical chemistry, synthesis, crystal growth and design, charge transfer complexes and ion-radical salts of fullerenes, air-sensitive compounds, conducting and magnetic properties, IR-, UV- VIS-NIR, and EPR spectroscopy, photoinduced electron transfer, phase transitions.

Professional Appointments:

Senior researcher, Institute of Problems of Chemical Physics RAS.

Awards and Honors:

Award of Journal Russian Chemical Review for most cited review: D.V. Konarev, R.N. Lyubovskaya, Russ. Chem. Reviews, 1999, 68(1), 19-38.

Publications during last three years:

1. [C60(CN)5]– : A remarkably stable [60]fullerene anion. P.A. Troshin, E.A. Khakina, A.S. Peregudov, D.V. Konarev, I.V. Soulimenkov, S.M. Peregudova, R.N. Lyubovskaya, Eur. J. Org. Chem., 3265-3268 (2010).
2. Outer-sphere coordination trinuclear assembly of manganese(II) tetraphenylporphyrins with octaazacrypt and stabilized in solid state by complex formation with fullerene C70. D.V. Konarev, S.S. Khasanov, R.N. Lyubovskaya, Cryst. Growth Des., 10, 1451-1454 (2010).
3. Magnetic and structural transitions at dimerization of C60·- in ionic fullerene complexes with metalloporphyrins: {M(II)TPP(TMP+)2}(C60-)2(C6H4Cl2)2 (C6H5 CN)2 (M = Zn and Mn). D.V. Konarev, S.S. Khasanov, G.R. Mukhamadieva, L.V. Zorina, A. Otsuka, H. Yamochi, R.N. Lyubovskaya, Inorg. Chem., 49, 3881-3887 (2010).
4. Unusual rotation and disorder-order transition in the molecular complex of fullerene C70 with covalently linked porphyrin dimer: {(Fe(III)TPP)2O}C70. D.V. Konarev, S.S. Khasanov, R.N. Lyubovskaya, J. Porph. Phth., 14, 293-297 (2010).
5. Two-dimensional organic metal based on fullerene. D.V. Konarev, S.S. Khasanov, A. Otsuka, M. Maesato, G. Saito, R.N. Lyubovskaya, Angew. Chem., 49, 4829-4832 (2010).
6. Ionic complexes of fullerene C70 with tetrakis(dimethylamonoethylene). The formation of diamagnetic singly bonded (C70-)2 dimers. D.V. Konarev, S.S. Khasanov, S.V. Simonov E.I. Yudanova, R.N. Lyubovskaya, CrystEngComm., 12, 3542 – 3546 (2010).
7. Ionic fullerene complex (DMI+)3(C60·-)(I-)2 with three-dimensional DMI+- I-network and hexagonal fullerene array. D.V. Konarev, S.S. Khasanov, A. Otsuka, G. Saito, R.N. Lyubovskaya, CrystEngComm., 11, 5, 811-816 (2009).
8. Design of molecular and ionic complexes of fullerene C60 with metal (II) octaethylporphyrins, MIIOEP, where M = Zn, Co, Fe and Mn by using coordination M-N(ligand) and M-C(C60-) bonds. D.V. Konarev, S.S. Khasanov, G. Saito, R.N. Lyubovskaya, Cryst. Growth Des., 9, 2, 1170-1181 (2009).
9. Effect of the Co-C(C60-) bond formation on magnetic properties of the ionic complex {Cryptand[2,2,2](Na+)}{Co(II)TPP(C60-)}(C6H4Cl2)2. D.V. Konarev, S.S. Khasanov, A. Otsuka, G. Saito, R.N. Lyubovskaya, Dalton Trans., 6416-6420 (2009).

10. Coordination structures of metalloporphyrins with N-containing cations in the complexes with fullerenes. D.V. Konarev, R.N. Lyubovskaya, Macroheterocycles, 2, 3-4, 191-197 (2009).
11. Neutral and ionic complexes of C60 with (ZnOEP)2.BPy coordination dimers. D.V. Konarev, S.S. Khasanov, Yu. L. Slovokhotov, G. Saito, R.N. Lyubovskaya, CrystEngComm., 10, 48-53. (2008).
12. Supramolecular approach in the synthesis of [60]fullerene – metal dithiocarbamate complexes, M(II)(dtc)2 (M= Zn, Cd, Hg, Fe and Mn). The study of their magnetic properties and photoconductivity. D.V. Konarev, A.Yu. Kovalevsky, S.S.Khasanov, D.V. Lopatin, A.V. Umrikhin, G. Saito, R.N.Lyubovskaya, Cryst. Growth Des., 8, 4, 1161-1172 (2008)
13. Multi-component ionic complex containing (MDABCO+)2.Mn(II)TPP assemblies and C60(2-) dianions: synthesis, crystal structure and magnetic properties. D.V. Konarev, S.S. Khasanov, G. Saito, R.N. Lyubovskaya, J. Porph. Phth., 12, 11, 1146-1153 (2008).
14. Magnetic properties of fullerene salts with d- and f-metal cations (Co(2+), Ni(2+), Fe(2+), Mn(2+), Eu(2+) and Cd(2+)). Specific features of the interaction between (60·-) and the metal cations. D.V. Konarev, R.N. Lyubobskaya, Russ. Chem. Bull., 57, 9, 1944-1954 (2008).

In: Chemistry Researcher Biographical Sketches
Editors: E. S. Huseynov and V. Babayev
ISBN: 978-1-62100-771-5

Chapter 90

BIOGRAPHICAL SKETCH

Name:	Title:	Date of Birth:
Inés Velasco	Doctor in Medicine. (University of Sevilla	08/03/1967

EDUCATION:

Institution and Location	Degree	Year Conferred	Scientific Field
Servicio de Ginecología y Obstetricia Hospital de Riotinto. Servicio Andaluz de Salud.			

CONTACT POINTS:

Avda. La Esquila 5. 21.660- Minas de Riotinto.
Huelva (Spain)

RESEARCH AND PROFESSIONAL EXPERIENCE:

Master Degree in Research Methodology.
Doctor in Medicine. (University of Sevilla)
Fetal Medicine Graduate (University of Barcelona)

Professional Appointments:

Iodine Deficiency Disorders and Thyroid Dysfunction Working Group.
Spanish Society of Obstetricians and Gynecologists.
Spanish Society of Endocrinology and Nutrition.

Publications during last three years:

Velasco López, I. Necesidades de yodo en el embarazo. Razones clínicas para evitar la deficiencia de yodo en el embarazo. Prog Obstet Ginecol 2007; 50 Supl. 2: 35-42.

Velasco López, I. Yodoprofilaxis en el embarazo. Una nueva mirada para una vieja historia.. Endocrinol Nutr 2007; 54 (Supl 3): 73-78.

Velasco López, I. Comentarios al artículo "Tirotoxicosis en una gestación inducida por yodoprofilaxis". Prog Obstet Ginecol 2008; 51(2): 109-110.

García-Fuentes E, Gallo M, García L, Prieto S, Alcaide-Torres J, Santiago P, Velasco I, Soriguer F. Amniotic fluid iodine concentrations do not vary in pregnant women with varying iodine intake. Br J Nutr 2008; 99: 1178-1181 (Factor de impacto: 2,764).

Velasco I, Naranjo S, López-Pedrera C, Garriga MJ, García-Fuentes E, Soriguer F. Use of povidone-iodine during the first trimester of pregnancy: a correct practice? BJOG 2009; 116: 452-455. (Factor de impacto:3,437).

Velasco I, Santiago P, Anarte MT, Soriguer F. Seguridad de la suplementación con yodo durante la gestación en regiones con déficit leve-moderado. Med Clin (Barc) 2009; 132(14): 565-567. (Factor de impacto:1,231).

Velasco I, Carreira M, Santiago P, Muela JA, García-Fuentes E, Sánchez-Muñoz B, Garriga MJ, Gónzalez-Fernández MC, Caballero FF, González-Romero S, Anarte MT, Soriguer F. Effect of iodine during pregnancy on neurocognitive development of children during the first two years of life J Clin Endocrinol Metab 2009 Sep; 94(9): 3234-41 (Factor de impacto:6,202).

elasco I, González-Romero S, Soriguer F. Iodine and thyroid function during pregnancy. Epidemiology 2010 May; 21 (3): 428-9. (Factor de impacto:5,511).

Santiago P, López JA, Velasco I, Sánchez B., Rodriguez A, Olmedo P, López-Jurado R, SánchezC, Casimiro-Soriguer F. Normal ranges of thyroid hormones in pregnant women. Journal of Maternal-Fetal & Neonatal Medicine 2010; 23 (Suppl 1), May: 586. (Factor de impacto: 1,362).

Grupo de trabajo sobre los trastornos relacionados con la deficiencia de yodo (TDY) de la Sociedad Española de Endocrinología y Nutrición (SEEN). Avances en la erradicación de la deficiencia de yodo en España. Endocrinol Nutr. 2010;57:87-9.

Santiago P, Berrío M, Olmedo P, Velasco I, Sánchez B, García Fuentes E, Martínez J, Casimiro-Soriguer F. Valores de referencia de hormonas tiroideas en la población de mujeres gestantes de Jaén. Endocrinol Nutr. 2011; 58(2): 62-67.

Vila L, Velasco I, González-Romero, Morales F, Sánchez E,. Cribado de la disfunción/autoinmunidad tiroideas en la población gestante: ¿Está justificado un cribado universal? Med Clin (Barc) 2011 (aceptado y en prensa) (Factor de impacto:1,362).

Soriguer F, Gutierrez-Repiso C, Gonzalez-Romero S, Olveira G, Garriga MJ, Velasco I, Santiago P, de Escobar GM, Garcia-Fuentes E; Iodine Deficiency Disorders Group of Spanish Society of Endocrinology and Nutrition.Iodine concentration in cow's milk and its relation with urinary iodine concentrations in the population. Clin Nutr 2011; 30: 44-48. (Factor de impacto: 3,274).

Soriguer F, Gutierrez-Repiso C, Rubio-Martín E, Linares F, Cardona I, López-Ojeda J, Pacheco M, Gonzalez-Romero S, Garriga MJ, Velasco I, Santiago P, Garcia-Fuentes E. Iodine intakes of 100-300 mg/d do not modify thyroid function and have modest anti-inflammatory effects. Br J Nutr 2011 (aceptado, en prensa) (Factor de impacto:3,446)

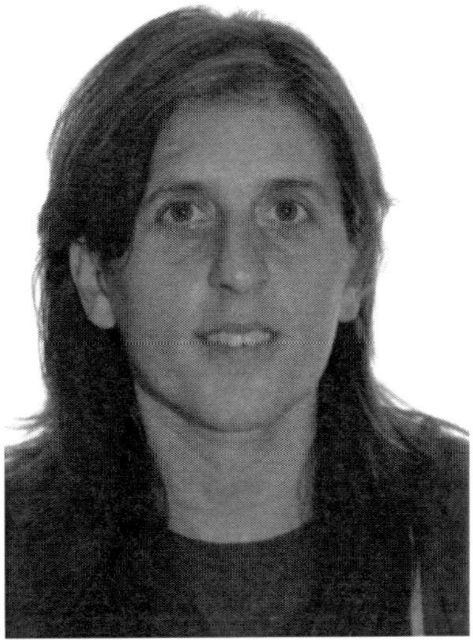

In: Chemistry Researcher Biographical Sketches
Editors: E. S. Huseynov and V. Babayev
ISBN: 978-1-62100-771-5

Chapter 91

BIOGRAPHICAL SKETCH

Name:	Title:	Date of Birth:
Gloria Víllora	University of Murcia	20/04/1957

EDUCATION:

Institution and Location	Degree	Year Conferred	Scientific Field
PhD Chemical Engineering			

CONTACT POINTS:

Department Chemical Engineering. Faculty of Chemistry. Campus de Espinardo. Universidad de Murcia. 30071 Murcia. Spain

RESEARCH AND PROFESSIONAL EXPERIENCE:

Professor of Chemical Engineering and Researcher

Professional Appointments:

Dean of the Faculty of Chemistry

Publications during last three years:

de los Ríos, A.P.; Hernández-Fernández, F.J.;Tomás Alonso, F.; Gómez, D., y Víllora G., "Biocatalytic kinetic resolution of *rac*-1-phenylethanol and *rac*-2-pentanol in hexane medium: acyl donor and water content effects". *Canadian Journal of Chemical Engineering. Vol. 88. Issue: 3, 442-446. June (2010).*

de los Ríos, A.P.; Hernández-Fernández, F.J.; Rubio M.; Gómez, D., y Víllora G., "Highly selective transport of transesterification reaction compounds through supported liquid membranes containing ionic liquids based on the tetrafluoroborate anion". *DESALINATION.* Vol. 250; 101-104 (2010).

Hernández-Fernández, F.J.; de los Ríos, A.P.; F.; Gómez, D., Rubio M. y Víllora G., "Selective extraction of organic compounds from transesterification reaction mixtures by using ionic liquids", *AiChE, Vol. 56, 1213-1217. May (2010).*

Hernández-Fernández, F.J.; de los Ríos, A.P.; Tomás, F.; Palacios, J.M. y Víllora G. "Preparation of supported ionic liquid membranes: Influence of the immobilization method of the ionic liquid on their operational stability". *JOURNAL OF MEMBRANE SCIENCE. Vol. 341, 172-177 (2009).*

Hernández-Fernández, F.J.; de los Ríos, A.P.; Tomás, F.; Palacios, J.M. y Víllora G. "Stability studies of supported liquid membranes based on ionic liquids: Effect of surrounding phase nature". *DESALINATION. Vol. 245, 776-782 (2009).*

Hernández-Fernández, F.J.; de los Ríos, A.P.; Tomás, F.; Gómez, D. y Víllora G. "Stability of hydrolase enzymes in ionic liquids". *The Canadian Journal of Chemical Engineering, Vol. 87, 910-914, Dec. (2009).*

Perez de los Rios, A.; Hernández Fernández, F. J.; Presa Garcia, H.; Gomez Perez, D.; Villora Cano, G. "Tayloring Supported Ionic Liquid Membranes For the Selective Separation of Transesterification Reaction Compounds". *Journal of Membrane Science 3.442, Vol. 328,81-85 (2009)*

Perez de los Rios, A.; Hernández Fernández, F. J.; Tomas Alonso, F.; Gomez Perez, D.; Villora Cano, G. "Improvement in the Separation Efficiency of Transesterification Reaction Compounds by the use of Supported Ionic Liquid Membranes Based on the Dicyanamide". *Desalination . Vol. 244,122-129 (2009)*

Perez de los Rios, A.; Hernández Fernández, F. J.; Tomas Alonso, F.; Rubio Torres, M.; Villora Cano, G. "Biocatalytic Kinetic Resolution of Rac-1-Phenylethanol and Rac-2-Pentanol In Hexane Medium: Acyl Donor And Water Content Effects". *Catalysis Today 3.108, Vol. 1,1-12 (2008)*

Hernández Fernández, F. J.; Perez de Los Rios, A.; Tomas Alonso, F.; Gomez Perez, D.; Villora Cano, G. "On the Development of an Integrated Membrane Process with Ionic Liquids for de Kinetic Resolution of Rac-2-Pentanol". *Journal of Membrane Science 3.442, Vol. 314,238-246 (2008)*

Perez de los Rios, A.; Hernández Fernández, F. J.; Tomas Alonso, F.; Rubio Torres, M.; Gomez Perez, D.; Villora Cano, G. "On the Importance of the Nature of the Ionic Liquids in the Selective Simultaneous Separation of the Substrates and Products of a Transesterification Reaction Through Supported Ionic Liquid Membranas". *Journal of Membrane Science 3.442, Vol. 307,233-238 (2008)*

Perez de los Rios, A.; Hernández Fernández, F. J.; Rubio Torres, M.; Tomas Alonso, F.; Gomez Perez, D.; Villora Cano, G. "Prediction of the Selectivity in the Recovery of Transesterification Reaction Products Using Supported Liquid Membranes Based on Ionic Liquids". *Journal of Membrane Science 3.442, Vol. 307,225-232 (2008)*

Perez de los Rios, A.; Hernández Fernández, F. J.; Tomas Alonso, F.; Palacios Latasa, J. M.; Villora Cano, G. "Stability Studies of Supported Liquid Membranes Based on Ionic Liquids: Effects of Surrounding Phase Nature". *Desalination .955, Vol. 11,20-40 (2008)*

Perez de los Rios, A.; Hernández Fernández, F. J.; Tomas Alonso, F.; Gomez Perez, D.; Rubio Torres, M.; Villora Cano, G. "Synthesis of Esters in Ionic Liquids. The Effect of Vinyl Esters and Alcohols". *Process Biochemistry 2.336, Vol. 43, Nº 8,892-895 (2008)*

Perez de los Rios, A.; Hernández Fernández, F. J.; Tomas Alonso, F.; Gomez Perez, D.; Villora Cano, G. "Synthesis of Flavour Esters Using Free Candida Antarctica Lipase B in Ionic Liquids". *Flavour and Fragrance Journal .628, Vol. 23,319-322 (2008)*

Antonia P. de los Ríos, Francisco J. Hernández-Fernández , Demetrio Gómez, F. Tomás-Alonso, Manuel Rubio and Gloria Víllora. "Ternary liquid-liquid equilibria for mixtures of an ionic liquid + hexane + an organic compound involved in the kinetic resolution of rac-1-phenyl ethanol (rac-1-phenyl ethanol, vinyl propionate, rac-1-phenylethyl propionate or propionic acid) at 298.2K and atmospheric pressure". *Fluid Phase Equilibria* 263, 190-198 (2008).

In: Chemistry Researcher Biographical Sketches
Editors: E. S. Huseynov and V. Babayev
ISBN: 978-1-62100-771-5

Chapter 92

BIOGRAPHICAL SKETCH

Name:	Title:	Date of Birth:
Maja D. Vitoroviæ-Todoroviæ	Military-Technical Institute	30.09.1977

EDUCATION:

Institution and Location	Degree	Year Conferred	Scientific Field
MSc. in Biochemistry (Faculty of Chemistry, University of Belgrade) PhD in progress, (Faculty of Chemistry, University of			

CONTACT POINTS:

Ratka Resanoviæa 1, 11000 Belgrade, Serbia (work)
III Nova 58, Pinosava, 11000 Belgrade, Serbia (home)

RESEARCH AND PROFESSIONAL EXPERIENCE:

Maja Vitoroviæ-Todoroviæ has graduated from Faculty of Chemistry, University of Belgrade (department of inorganic chemistry) in year 2004., with thesis “Synthesis and characterization of Zn(II), Cd (II) and Ru(III) complexes with N-carbobenzyloxy-glycine”. She finished her MSc thesis “4-aryl-4-оxо-N-phenyl-2-aminylbutyramides – synthesis, anticholinesterase activity and 3D-QSAR” in year 2010, at Faculty of Chemistry, University of Belgrade at Department of Biochemistry. Now, she is doing her PhD thesis also at Faculty of Chemistry, University of Belgrade. Her research interests are: - Medicinal chemistry and enzymology. Design, synthesis, purification and structural characterization of organic compounds and evaluation of their biological activity (anticholinesterase, anti-tumor, protective effects against OP irreversible inhibition). Interactions of small organic molecules with their respective biological targets. Molecular modeling techniques: docking, 3D quantitative structure activity relationships and molecular dynamic simulations. -Application of metal oxides nanopartcles (MONP’s) with self decontamination

In: Chemistry Researcher Biographical Sketches
Editors: E. S. Huseynov and V. Babayev
ISBN: 978-1-62100-771-5

Chapter 93

BIOGRAPHICAL SKETCH

Name:	Title:	Date of Birth:
Vladimir Volodin	A.V. Rzhanov Institute of Semiconductor Physics SB RAS	31 July 1967

EDUCATION:

Institution and Location	Degree	Year Conferred	Scientific Field
Graduate project in Novosibirsk State University, Physics Department, Chair of Physics of semiconductors, Subject – Studies of radiation and thermo-stimulated stress relaxation in silicon on sapphire structures.\		1991	
Ph.D. thesis, Subject – Raman scattering study of Si and GaAs based semiconductor nanostructures		1999	
Diploma of Professor-Assistance in area of Semiconductor physics		2005	

CONTACT POINTS:

Lavrent'eva ave., 13, 630090, Novosibirsk, Russia

RESEARCH AND PROFESSIONAL EXPERIENCE:

1984-1986 - Student of Novosibirsk State University (NSU), Physics Department.
1986-1988 - Private (soldier) in Soviet Army.
1988-1991 - Student of NSU, Physics Department, specialized on Chair of Physics of Semiconductors.
1991-1993 - Researcher-engineer in Institute of Semiconductor Physics (ISP) of Siberian Branch of Russian Academy of Sciences (SB RAS)
1993-1997 - Junior-researcher in ISP SB RAS
1997-2001 - Researcher, group leader in ISP SB RAS
2001-present time - Senior researcher, group leader in ISP SB RAS

Professional Appointments:

Generally, I study semiconductor heterostructures using optical techniques (Raman scattering, ellipsometry, spectroscopy of transmission, photoluminescence). My PhD thesis was devoted to Raman study of semiconductor nanoobjects (nanocrystals, quantum well wires and quantum dots). In present time I am interested in study of silicon nanocrystals in a-Si, SiO_2 and SiN_x matrixes and germanium nanocrystals in GeO_2 matrix. I am also interested in study of phonon and coupled phonon-plasmon and phonon-electron modes in tunneling semiconductor structures – doped silicon nanocrystals, GaAs/AlAs superlattices and quantum well wires (QWWs) and also in study of QWWs and quantum dots formation on high-index and stressed surfaces.

Also, during 20 last years I dealt with pulse laser annealing of amorphous silicon films and semiconductor nanostructures. Last time in addition to nanosecond laser crystallization I am interesting in femtosecond laser crystallization and not thermal effects in phase transitions.

Awards and Honors:

1990-Diploma at All-Soviet Union Student Conference (section of the Physics of Semiconductor).
1997- Grant of SB RAS for young scientists.
1996-2003- Fellowship of Scientific Council of ISP SB RAS for Young Scientists.

Publications during last three years:

1. V.A. Volodin, E.B. Gorokhov, D.V. Marin, H.Rinnert, P.Miska, M.Vergnat. Quasi-direct optical transitions in Ge nanocrystals embedded in GeO_2 matrix. JETP Letters, Vol. 89, No. 2, pp. 76–79, (2009).
2. D.V. Marin, E.B. Gorokhov, A.G. Borisov, V.A. Volodin. Ellipsometry of GeO_2 Films with Ge nanoclusters: Influence of the Quantum-Size Effect on Refractive Index.

Optics and Spectroscopy (Russian - Optika i Spektroskopiya), Vol. 106, No. 3, pp. 436–440, (2009).

3. I.V. Antonova, M.B. Gulyaev, A.G. Cherkov, V.A. Volodin, D.V. Marin, V.A. Skuratov, J.Jedrzejewski, and I. Balberg. The modification of Si nanocrystallites embedded in a dielectric matrix by high energy ion irradiation. Nanotechnology, v. 20, n. 9, pp. 095205-(1-5), (2009).
4. T. T. Korchagina, D. V. Marin, V. A. Volodin, A. A. Popov, and M. Vergnat. Structure and Optical Properties of SiNx:H Films with Si Nanoclusters Produced by Low-Frequency Plasma-Enhanced Chemical Vapor Deposition. Semiconductors, Vol. 43, No. 11, pp. 1514–1520 (2009).
5. Irina V. Antonova, Vladimir A. Volodin, Efim P. Neustroev, Svetlana A. Smagulova, Jedrzej Jedrzejewsi, and Isaac Balberg. Charge spectroscopy of Si nanocrystallites embedded in SiO_2 matrix. JOURNAL OF APPLIED PHYSICS 106, 064306 (2009).
6. V.A. Volodin. Experimental Discovery of the Anisotropy of Phonon–Plasmon Modes in GaAs/AlAs(100) Superlattices JETP Letters, Vol. 89, No. 8, pp. 483–485, (2009).
7. V. Volodin, V. Stuchinsky, S. Cherkova, G. Kachurin, and R. Yankov. Temperature-Dependent PL Intensity in FLA-Synthesized and Hydrogen-Modified Silica Layers with Silicon Nanocrystals. ECS Trans. 16 (40), 105 (2009).
8. V. Volodin, V. Stuchinsky, T. Korchagina, A. Popov, and M. Vergnat. Temperature Dependence of the Photoluminescence Intensity in $Si_{3+x}N_4$:H Films with Amorphous Si Nanoclusters: Evidence for Two Processes Involved in the Nonradiative Relaxation of Photoexcitations. ECS Trans. 25 (11), 35 (2009).
9. V.A. Volodin, T.T. Korchagina, J. Koch, B.N. Chichkov. Femtosecond laser induced formation of Si nanocrystals and amorphous Si clusters in silicon-rich nitride films. Physica E, v.42, n.6, pp 1820-1823 (2010).
10. D.V. Marin, V.A. Volodin, E.B. Gorokhov, D.V. Shcheglov, A.V. Latyshev, M. Vergnat, J. Koch, B.N. Chichkov. Modification of Germanium Nanoclusters in GeOx Films during Isochronous Furnace and Pulse Laser Annealing. Technical Physics Letters, Vol. 36, No. 5, pp. 439–442 (2010).

In: Chemistry Researcher Biographical Sketches
Editors: E. S. Huseynov and V. Babayev
ISBN: 978-1-62100-771-5

Chapter 94

BIOGRAPHICAL SKETCH

Name:	Title:	Date of Birth:
Winfried Vonau	Prof. Dr.	19/12/1956

EDUCATION:

Institution and Location	Degree	Year Conferred	Scientific Field
University of Leipzig, Germany			

CONTACT POINTS:

Kurt-Schwabe-Institut für Mess- und Sensortechnik e.V. Meinsberg/ Kurt-Schwabe-Straße 4/ D-04720 Ziegra-Knobelsdorf, Germany

RESEARCH AND PROFESSIONAL EXPERIENCE:

Electrochemical Sensors and Analytical Chemistry

Professional Appointments:

Deputy Director of the Kurt-Schwabe-Institute

Publications during last three years:

Herrmann, S.; Berthold, F.; Gerlach, F.; Vonau, W.; Mayer, M.; Bieger, W.; Klare. P.: Chip-Messsysteme für medizinische Applikationen. Mikrosystemtechnik Chemnitz 07, 8. Chemnitzer Fachtagung Mikromechanik & Mikroelektronik. Chemnitz, 14.-15.11.2007, Tagungsband S. 125-128

Enseleit, U. Bachmann, T. Spindler, J. Gerlach, F. Vonau, W.: Beispiele zur Realisierung gemischtleitender Ableitsysteme im Kontakt mit kationenselektiven Gläsern. In: G. Gerlach, P. Hauptmann (Hg.): Dresdner Beiträge zur Sensorik, Band 29: 8. Dresdner Sensor-Symposium – Sensoren für Umwelt, Klima und Sicherheit; Biosensoren und Biosysteme; Sensoren und Sensorsysteme für die Prozesstechnik; Trends in der Sensortechnik; Materialentwicklungen für die Sensorik. Dresden, TUDpress 2007,S. 291-294

Jahn, H.; Herrmann, S.; Gerlach, F.; Köhler, J.; Vonau, W.: Mehrparametermesssystem für biotechnologische Prozesse. In: G. Gerlach, P. Hauptmann (Hg.): Dresdner Beiträge zur Sensorik, Band 29: 8. Dresdner Sensor-Symposium – Sensoren für Umwelt, Klima und Sicherheit; Biosensoren und Biosysteme; Sensoren und Sensorsysteme für die Prozesstechnik; Trends in der Sensortechnik; Materialentwicklungen für die Sensorik. Dresden, TUDpress 2007,S. 241-244

Gerlach, F.; Seitz, D.; Herrmann, S.; Mayr, H.; Ziegler, G.; W. Vonau, W.: Konzeption für ein kompatibles Sensorarray-System mit Schnellankopplungs-mechanismus für eine Online-Messung in der Zellkulturtechnik. In: G. Gerlach, P. Hauptmann (Hg.): Dresdner Beiträge zur Sensorik, Band 29: 8. Dresdner Sensor-Symposium – Sensoren für Umwelt, Klima und Sicherheit; Biosensoren und Biosysteme; Sensoren und Sensorsysteme für die Prozesstechnik; Trends in der Sensortechnik; Materialentwicklungen für die Sensorik. Dresden, TUDpress 2007,S. 245-248

Herrmann, S.; Berthold, F.; Mayer, Bieger, W.; Vonau, W.: Sensormodul zur Bestimmung von Life Sciences relevanten Parametern.In: G. Gerlach, P. Hauptmann (Hg.): Dresdner Beiträge zur Sensorik, Band 29: 8. Dresdner Sensor-Symposium – Sensoren für Umwelt, Klima und Sicherheit; Biosensoren und Biosysteme; Sensoren und Sensorsysteme für die Prozesstechnik; Trends in der Sensortechnik; Materialentwicklungen für die Sensorik. Dresden, TUDpress 2007,S. 175-178

Hüttl, R.; Harmel, J.; Lissner, A.; Wolf, G.; Klare, P.; Vonau, W.; Berthold, F.; Herrmann, S.: A small-scale calorimetric reactor system combined with several chemical sensors for the investigation of microbial growth processes. Engineering in Life Sciences 8(1) (2008) 56-61.

Vonau, W.; Herrmann, S.;Berthold, F.; Mayer, W.; Bieger, W.: Sensorsignale aus dem Blut - Multiparameter-Messsystem. Chemie Ingenieur Technik 2007, 79. No. 9, 1385

Seitz, D.; Gerlach, F.; Herrmann, S.; Mayr, H.; Vonau, W.; Ziegler, G.: A biocompatible sensor-array instrument with rapid coupling mechanism for online measurement of physiological data in cell cultures. In: Regenerative Medicine 2 (5) 2007 686

Vonau, W.; Oelßner, W.; Zosel, J.; Henze, J.; Guth, U.: Elektrochemische Festkörper-Kombinationselektrode für Anwendungen in der Lebensmittelindustrie. Chemie Ingenieur Technik 2007, 79, No. 9, 1446

Guth, U.; Vonau, W.: pH-Wert-Messung in ausgewählten Applikationsgebieten – keine Routineaufgabe! Mitteldeutsche Mitteilungen 1/2008, S. 20

W. Vonau; Gerlach, F.; Enseleit, U.; Spindler J.; Bachmann, T.: New solid-state glass electrodes by using zinc oxide thin films as interface layer. Journal of Solid State Electrochemistry. 13 (2009) 91-98

Guth, U.; Gerlach, F.; Decker, M.; Oelßner W.; Vonau, W.: Solid-state reference elec-trodes for potentiometric sensors. Journal of Solid State Electrochemistry. 13 (2009) 27-39

Guth, U.; Vonau, W.; Zosel, J.; Recent developments in electrochemical sensor appli-cation and technology - a review. Measurement Science and Technology. Meas. Sci. technol. 20 (2009) 042002 (14pp)

Vonau, W.; Enseleit, U.; Gerlach, F.: Multi-flow-through sensor for simultaneous cation determination. Electrochemistry Communications 10 (2008) 1355-135

Herrmann, S.; Gerlach, F.; Seitz, D.; Deisinger, U.; Ziegler, G.; Vonau, W.: Sensorarray-System mit Schnellankopplungsmechanismus zur Online-Messung in der Zellkulturtechnik. 14. Heiligenstädter Kolloquium. Tagungsband (2008) S. 405-413

Hüttl, R.; Harmel, J.; Lissner, A.; Wolf, G.; Klare, P.; Vonau, W.; Berthold, F.; Herrmann, S.: Miniaturisiertes kalorimetrisches Reaktorsystem mit integrierter Fluidik und Sensorik. 14. Heiligenstädter Kolloquium. Tagungsband (2008) S. 195-199

Seitz, D.; Gerlach, F.; Herrmann, S.; Vonau, W.; Ziegler, G.: Online-Kontrolle biologi-scher Parameter unterhalb von adhärenten Zellen und 3D-Konstrukten mit elektroche-mischen Sensoren. In: R. Poll, J. Füssel, E. Koch (Hg.): Dresdner Beiträge zur Medizin-technik, Band 6:2. Dresdner Medizintechnik-Symposium – Innovation durch Einheit von Therapie und Monitoring. Dresden, TUDpress 2008, S. 65-67

Vonau, W.; Gerlach, F.; Herrmann, S.; Seitz, D.; Ziegler, G.: Tools zur Charakterisierung des Kalibrationsprozesses adhärenter, humaner Zellen. In: R. Poll, J. Füssel, E. Koch (Hg.): Dresdner Beiträge zur Medizintechnik, Band 6: 2. Dresdner Medizintechnik-Symposium – Innovation durch Einheit von Therapie und Monitoring. Dresden, TUDpress 2008, S. 182-187

Vonau, W.; Oelßner, W.; Guth, Henze, J.: An all-solid-state reference electrode. Sensors and Actuators B 144 (2010) 368–373

W. Vonau, W.; U. Enseleit, U.: Ionenselektive Feststoffsensoren mit amorpher Membran. GIT Labor-Fachzeitschrift 10 (2009) 664-666

Enseleit, U.; Decker, M., Vonau, W.: Neuartige Sensorelektrolytsysteme für die dCO2-Bestimmung bei höheren Kohlendioxiddrucken in Bioreaktoren. In: G. Gerlach, P. Hauptmann (Hg.): Dresdner Beiträge zur Sensorik, Band 39: 9. Dresdner Sensor-Symposium. Dresden, TUDpress 2009, S. 189-192

Vonau, W.; Zosel, J., Oelßner, W.: Editorial. Technisches Messen 77 (2010) 3-4

Vonau, W.; Oelßner, W.; Zosel, J.; Kohler, H.: Die Bedeutung der Materialentwicklung für elektrochemische Sensoren. Technisches Messen 77 (2010) 5-18

Päßler, S.; Herrmann, S.; Vonau, W.; Bertau, M.: Optimierung bioverfahrenstechnicher Prozesse mittels sensorgestütztem Mikrobioreaktor. Technisches Messen 1 (2010) 24-29

Vonau, W.; Berthold, F.; Jahn, H.; Guth, U.: Long-term investigation of glass electrodes at higher temperatures. Chem. Anal. (Warzawa) 54 (2009) 1181-1192

Vonau, W.; Herrmann, S.; Gerlach, F.: Antimony electrodes for pH measurements in special applications. Chem. Anal. (Warzawa) 54 (2009) 1169-1179

Vonau, W.: Elektrochemische pH-Sensoren für Spezialanwendungen. Technisches Messen 77 (2010) 162-172.

Vonau, W.; Gerlach, F.; Herrmann, S.: Conception of a new technique in cell cultivation using a lab-on-chip aided miniaturised device with calibratable electrochemical sensors. Microchim. Acta doi 10.1007/s00604-010-0427-9 (in press)

Vonau, W.; Berthold, F.; Guth, U.: Impedance investigations on heat-stressed, pH-sensitive glass membranes. Journal of Solid State Electrochemistry. doi 10.1007/s10008-010-1148-z (in press)

In: Chemistry Researcher Biographical Sketches
Editors: E. S. Huseynov and V. Babayev
ISBN: 978-1-62100-771-5

Chapter 95

BIOGRAPHICAL SKETCH

Name:	Title:	Date of Birth:
Andriy Voronov	Coatings and Polymeric Materials Department, North Dakota State University	July, 21, 1968

EDUCATION:

Institution and Location	Degree	Year Conferred	Scientific Field
Lviv Polytechnic National University of Ukraine Chemical Engineering.	M.S	1990	
Lviv Polytechnic National University of Ukraine Macromolecular Chemistry	Ph.D.,	1994	

CONTACT POINTS:

Dep.2760, PO Box 6050, Fargo, ND, USA 58108-6050

RESEARCH AND PROFESSIONAL EXPERIENCE:

2000-2003 - Alexander von Humboldt-Fellow, Department of Physical Chemistry, University of Bayreuth, Germany

1998-1999 – Research Scientist, Department of Experimental Physics, University of Ulm, Germany

1997-1998 - Research Scientist, Department of Colloids and Interfaces, Institute Charles Sadron, CNRS, Strasbourg, France

Professional Appointments:

Nov. 2007 – present Assistant Professor, Department of Coatings and Polymeric Materials, North Dakota State University, Fargo, ND

2003 – 2007 - Staff Scientist, Institute of Particle Technology, University of Erlangen-Nuremberg, Germany

1999-2000 - Lecturer, Lviv Polytechnic National University of Ukraine

1990-1996 - Research Fellow, Institute of Physical Chemistry, NAS, Ukraine

Awards and Honors:

3M Non-tenured faculty awarded 2011

Long-State Fellowship awarded by Alexander von Humboldt Fellowship for the research at the University of Bayreuth, Germany, 2001.

Alexander von Humboldt Fellowship awarded by Alexander von Humboldt Foundation for the research at the University of Bayreuth, Germany, 2000.

Research Fellowship awarded by NATO for the research at Institute Charles Sadron, CNRS, Strasbourg, France 1997.

Faculty Research Fellow Award awarded by Austrian Ministry of Education, 1995.

Publications during last three years:

Hevus, I., Kohut, A., Voronov, A. (2010) Amphiphilic invertible polyurethanes: synthesis and properties. Macromolecules, 43(18), 7488-7494.

Sieburg, L., Kohut, A., Kislenko, V., Voronov, A. (2010) Amphiphilic invertible polymers for adsolubilization on hydrophilic and hydrophobized silica nanoparticles. J. Colloid Interface Sci. 351 (1), 116–121.

Varvarenko, S.; Voronov, A.; Samaryk, V.; Tarnavchyk, I.; Nosova, N.; Kohut, A.; Voronov, S. (2010) Covalent grafting of polyacrylamide-based hydrogels to a polypropylene surface activated with functional polyperoxide modification React. Funct. Polym. 70 (9), 647–655.

Tarnavchyk I., Voronov A., Kohut A., Nosova N., Varvarenko S., Samaryk V., Voronov S. (2009) Reactive Hydrogel Networks for Fabrication of Metal-Polymer Nanocomposites. Macrom.Rapid Comm., 30 (18), 1564 – 1569.

Samaryk, V.; Tarnavchyk, I.; Voronov, A.; Varvarenko, S.; Nosova, N.; Kohut, A.; Voronov, S. (2009) A New Acrylamide-Based Peroxide Monomer: Synthesis and Copolymerization with Octyl Methacrylate Macromolecules, 42 (17), 6495–6500.

Kohut A, Voronov A. (2009) Hierarchical Micellar Structures from Amphiphilic Invertible Polyesters: ^{1}H NMR Spectroscopic Study. Langmuir, 25, 4356-4360.

Shafranska O., Voronov A., Kohut A., Xiang-Fa Wu, Akhatov I. (2009) Polymer-metal complexes . Carbon, 47 (13), 3137 – 3139.

Voronov A., Kohut, A., Vasylyev S., Peukert W. (2008) Mechanism of Silver Ion Reduction in Amphiphilic Invertible Nanoreactors formed in a Non-polar Solvent at Ambient Conditions, Langmuir, 24 (21), 12587-12594.

Martinez Tomalino, L., Voronov, A. Kohut,A., Peukert,W. (2008) Study of Amphiphilic Polyester Micelles by Hyper-Rayleigh Scattering: Invertibility and Phase Transfer, Journal of Physical Chemistry, 112, 6338 - 6343.

In: Chemistry Researcher Biographical Sketches
Editors: E. S. Huseynov and V. Babayev
ISBN: 978-1-62100-771-5

Chapter 96

BIOGRAPHICAL SKETCH

Name:	Title:	Date of Birth:
Viroj Wiwanitkit	Professor, M.D	1975

CONTACT POINTS:

Wiwanitkit House, Bangkhae, Bangkok Thailand; visiting university professor, Hainan Medical University, China

RESEARCH AND PROFESSIONAL EXPERIENCE:

More than 850 international publications

Publications during last three years:

More than 850 international publications.

In: Chemistry Researcher Biographical Sketches
Editors: E. S. Huseynov and V. Babayev

ISBN: 978-1-62100-771-5

Chapter 97

BIOGRAPHICAL SKETCH

Name:	Title:	Date of Birth:
Anatoly G. Zakharov	Institute of Solution Chemistry of the Russian Academy of Sciences	09.05.1946

EDUCATION:

Institution and Location	Degree	Year Conferred	Scientific Field
University of Chemistry and Technology, Ivanovo			

CONTACT POINTS:

Akademicheskaya St., 1, Ivanovo, 153045, Russia

RESEARCH AND PROFESSIONAL EXPERIENCE:

Thermodynamics of polymer-liquids systems

Professional Appointments:

Director, head of a department

Awards and Honors:

Professor

Publications during last three years:

1. A.G. Zakharov, M.I.Voronova, O.V. Surov, D.V. Batov Phenol sorption on cellulose from binary aqueous-organic mixtures// J Molecular Liquid, 2010, V.151, N1, pp.74-80.
2. O. V. Surov, M. I. Voronova, N. Zh. Mamardashvili, A. G. Zakharov Vapor pressures of macrocyclic compounds according to effusion method data// Tetrahedron Letters, 2011, 52, pp. 705-707.
3. A.G. Zakharov, M.I.Voronova, , D.V. Batov, K.V. Smirnova Thermochemical Characteristics of Solution of Phenol and Benzoic Acid in Water-Dimethylsulfoxide and Water-Acetonitrile Mixtures// Russian Journal of Physical Chemistry A, 2011, Vol. 85, No. 3, pp. 408-412.
4. O. V. Surov, M. I. Voronova, N. Zh. Mamardashvili, and A. G. Zakharov Thermodynamics of Solvation of Calix[4]arenes in n-Hexane// Russian Journal of Physical Chemistry A, 2010, Vol. 84, No. 11, pp. 1867–1872.
5. M.I. Voronova, A.G. Zakharov Adsorption of Phenol and Toluene from the gas phase and aqueous solutions on cellulose// Russian Journal of Applied Chemistry, 2009, V.82, N3, p.402-405.
6. O. V. Surov, M. I. Voronova, P. R. Smirnov, V. N. Trostin, N. Zh. Mamardashvili,and A. G. Zakharov Polymorphic Conversions of 4-tert-Butylcalix[4]arene upon the Formation and Thermal Destruction of a Complex with n-Hexane// Russian Journal of Physical Chemistry A, 2011, Vol. 85, No. 7, pp. 1162–1167
7. M.I. Voronova, D.A. Kuznetsov, A.G. Zakharov Sorption of Phenol on cellulose from binary aqueous-organic mixtures// Russian Journal of Applied Chemistry, 2008, V.81, N11, p.1924-1927.
8. M. I. Voronova, O. V. Surov, D. V. Batov, and A. G. Zakharov The Adsorption of Aqueous–Organic Mixture Components on Cellulose//Russian Journal of Physical Chemistry A, 2008, Vol. 82, No. 3, pp. 405–409.

In: Chemistry Researcher Biographical Sketches
Editors: E. S. Huseynov and V. Babayev

ISBN: 978-1-62100-771-5

Chapter 98

BIOGRAPHICAL SKETCH

Name:	Title:	Date of Birth:
Aristides D. Zdetsis	Professor, Dr	May 31-1946

EDUCATION:

Institution and Location	Degree	Year Conferred	Scientific Field
Physics, University of Athens Greece	B.S		

CONTACT POINTS:

zdetsis@upatras.gr , zdetsis@gmail.com
URL:http://moleng.physics

RESEARCH AND PROFESSIONAL EXPERIENCE:

Publications during last three years:

“Theoretical Predictions of a New Family of Stable Bismuth and Other Group 15 Fullerenes”, Aristides D. Zdetsis,J. Chem. Phys. C 114, 10775 (2010)

“Structural reciprocity effect in binary silicon-bismuth clusters”, Aristides D. Zdetsis, Chem. Phys. Lett. 493, 45 (2010)

“Rationalizing and Functionalizing Stannaspherene: Very Stable Stannaspherene “Alloys””, J. Chem. Phys. 131, 224310 (2009)

“Stabilization of large silicon fullerenes and related nanostructures trough puckering and poly(oligo)merization”, A. D. Zdetsis, Phys. Rev. B 80, 195417, selected for the November 30, 2009 issue of AIP's Virtual Journal of Nanoscale Science & Technology. (2009)

"Silicon-Bismuth and Germanium-Bismuth Clusters of High Stabilty", A. D. Zdetsis, J. Phys. Chem. A 113, 12079-12087 (2009)

"Multidecker Sandwiches of Silicon-Carbon Clusters", E. N. Koukaras and A. D. Zdetsis, Organometallics 28 (15), 4308–4315 (2009).

"One-nanometer luminous silicon nanoparticles: Possibility of a fullerene interpretation", Aristides D. Zdetsis,Phys. Rev. B 79, 195437 (2009), selected for the June 8, 2009 issue of AIP's Virtual Journal of Nanoscale Science & Technology.

"A parallel study of Ni@Si12 and Cu@Si12 nanoclusters", A. D. Zdetsis, E. N. Koukaras and C. S. Garoufalis, J. Math. Chem. 46, 971 (2009) "Mixed silicon-germanium nanocrystals: A detailed study of SixGe47-x:H", A. D. Zdetsis, C. S. Garoufalis and E. N. Koukaras, J. Math. Chem. 46, 942 (2009)

"Optical properties of ultra small Si nanoparticles: potential role of surface reconstruction and oxygen contamination", A. D. Zdetsis and C. S. Garoufalis, J. Math. Chem. 46, 952 (2009)

"Properties of silicon nanoparticles via a transferable tight-binding Hamiltonian based on ab initio results", A. D. Zdetsis and N. C. Bacalis, J. Math. Chem. 46, 962 (2009) "Structural and electronic properties of semiconductor clusters", Aristides D. Zdetsis, AIP Conf. Proc. 1148, 367 (2009)

"A systematic way for obtaining the structural and electronic properties of Silicon-Carbon clusters of the form SinC2, n=1–5", E. N. Koukaras, M. Mühlhäuser and A. D. Zdetsis, AIP Conf. Proc. 1148, 384 (2009)

"Multidecker stacking and cluster fusion of silicon-carbon clusters", E. N. Koukaras and A. D. Zdetsis, AIP Conf. Proc. 1148, 368 (2009)

"Elastic properties of ultra-thin hydrogenated silicon nanowires based on all electron mixed ab initio and semiempirical calculations", E. N. Koukaras, C. S. Garoufalis and A. D. Zdetsis, AIP Conf. Proc. 1148, 376 (2009)

"Success and pitfalls of the Sin 2C2H2–C2Bn 2Hn isolobal analogy: Depth and breadth of the boron connection", Aristides D. Zdetsis, J. Chem. Phys. 130, 064303 (2009)

"From "structural democracy" to "boron connection"", Aristides D. Zdetsis, AIP Conf. Proc. 1108, 231 (2009)

"Invited Paper. The Boron Connection: From Si6 to C6 and Much More", Aristides D. Zdetsis, Comp. Lett. 4, 1 (2008) "Aromaticity of planar Si6 rings in silicon-lithium clusters", Aristides D. Zdetsis, Patrick W. Fowler and Remco W. A. Havenith, Mol. Phys. 106, 1803, (2008)

"Novel effects in finite-length silicon nanowires", A. D. Zdetsis, E. N. Koukaras and C. S. Garoufalis, Phys. Stat. Sol. (a) 205, 2625, (2008)

"The boron connection: a parallel description of aromatic, bonding, and structural characteristics of hydrogenated silicon-carbon clusters and isovalent carboranes", Aristides D. Zdetsis, Inorg. Chem. 47, 8823, (2008)

"A new class of silicon-carbon clusters: A full study of the hydrogenated SinC2H2, n=3,4,5, clusters in comparison with their isoelectronic carboranes C2BnHn+2", A. D. Zdetsis, J. Chem. Phys. 128, 184305, (2008), selected for the May 26, 2008 issue of AIP's Virtual Journal of Nanoscale Science & Technology.

"High-Stability Hydrogenated Silicon Carbon Clusters: A Full Study of Si2C2H2 in Comparison to Si2C2, C2B2H4, And Other Similar Species", A. D. Zdetsis, J. Phys. Chem. A 112, 5712, (2008)

"High-symmetry low-energy structures of C60H60 and related fullerenes and nanotubes", A. D. Zdetsis, Phys. Rev. B 127, 115402, (2008).

"Analogy of silicon clusters with deltahedral boranes: How far can it go? Reexamining the structure of Sin and Sin2 , n=5–13 clusters", A. D. Zdetsis, J. Chem. Phys. 127, 244308, (2007).

"Stabilization of flat aromatic Si6 rings analogous to benzene: Ab initio theoretical prediction", A.D. Zdetsis, J. Chem. Phys. 127, 214306 (2007). "Similarities and Differences Between Silicon and Carbon Nanostructures: Theoretical Predictions", A.D. Zdetsis, AIP Conf. Proc. 963, 474 (2007).

"Electronic and Structural Properties of M@Si12 and M@Si12H12 Clusters, M = Ni, Zn ", E. N. Koukaras and A. D. Zdetsis, AIP Conf. Proc. 963, 395 (2007)

"Is Aromaticity and Fluxionality the Key to Magicity of Si6 Cluster?", A.D. Zdetsis, AIP Conf. Proc. 963, 383 (2007). "High-stability Finite-Length Silicon Nanowires: A Real Space Theoretical Study", E. N. Koukaras, A. D. Zdetsis and C. S. Garoufalis, AIP Conf. Proc. 963, 375 (2007).

"Symposium 2 Preface", A.D. Zdetsis, AIP Conf. Proc. 963, 357 (2007). "High-symmetry high-stability silicon fullerenes: A first-principles study", A.D. Zdetsis, Phys. Rev. B 76, 075402 (2007).

"Structural properties and magic structures in hydrogenated finite and infinite silicon nanowires", A. D. Zdetsis, E. N. Koukaras and C. S. Garoufalis, App. Phys. Lett. 91, 203112, (2007).

"Fluxional and aromatic behavior in small magic silicon clusters: A full ab initio study of Sin, Sin1-, Sin2-, and Sin1+, n=6, 10 clusters", A. D. Zdetsis, J. Chem. Phys. 127, 014314 (2007).

"Bonding and structural characteristics of Zn-, Cu-, and Ni-encapsulated Si clusters: Density-functional theory calculations", Aristides D. Zdetsis, Phys. Rev. B 75, 085409 (2007)

In: Chemistry Researcher Biographical Sketches
Editors: E. S. Huseynov and V. Babayev
ISBN: 978-1-62100-771-5

Chapter 99

BIOGRAPHICAL SKETCH

Name:	Title:	Date of Birth:
Jasenka Piljac Zegarac	Institute Rudjer Boskovic	April 29, 1976

EDUCATION:

Institution and Location	Degree	Year Conferred	Scientific Field
	BS Ph.D		Biochemistry, Biology

CONTACT POINTS:

Bijenicka cesta 54, 10000 Zagreb, Croatia

RESEARCH AND PROFESSIONAL EXPERIENCE:

Institute of Biological Chemistry, Washington State University, Pullman, WA, USA
Visiting Scientist ,August 2009 – current

Ruđer Bošković Institute, Zagreb, Croatia
Research Associate (permanent position, on leave, April, 2006 – current
Senior Research Assistant, March, 2004 – March, 2006

Research Assistant, Zagreb, Croatia; November, 1998 – October 2003

Faculty of Chemical Engineering and Technology, University of Zagreb, Croatia
Research/Teaching Assistant, Zagreb, Croatia; November, 1998 – October 2003

Awards and Honors:

Award	Institution	Year
Scientific Research Excellence Award	Ministry of Science, Education and Sports, Croatia	2008
Top 10% graduating class, UC Davis, Graduated with Honors	University of California, Davis	1997
California grant A	University of California, Davis	1996
Tracy and Ruth Storer Scholarship	University of California, Davis	1996/97
National Dean's List Award	University of California, Davis	1996
Tracy and Ruth Storer Scholarship	University of California, Davis	1995/96
UC Regents scholarship	University of California, Davis	1995/96
Novo Nordisk Biotech. Scholarship (awarded to high school seniors)	Novo Nordisk Biotech., Davis	1993

Publications during last three years:

- Šamec, D.; Piljac-Žegarac, J. *; Bogović, M.; Habjanič, K.; Grúz, J. (2011) Antioxidant potency of white (*Brassica oleracea* L. var. *capitata*) and Chinese (*Brassica rapa* L. var. *pekinensis* Lour.) cabbage: the influence of development stage, cultivar chioce and seed selection. *Scientia Horticulturae,*128(2): 78-83.
- Piljac-Žegarac, J. *; Šamec, D. (2011) Antioxidant stability of small fruits in postharvest storage at room and refrigerator temperatures. *Food Research International,* 44(1): 345-350.
- Belščak, A.; Bukovac, N.; Piljac-Žegarac, J. * (2011) The influence of ascorbic acid and honey addition on the antioxidant properties of fruit tea infusions. *Journal of Food Biochemistry* 35(1): 195-212.
- Zovko Končić, M.; Kremer, D.; Gruz, J.; Strnad M.; Biševac, G.; Kosalec, I.; Šamec, D.; Piljac-Žegarac, J.; Karlović, K. (2010) Antioxidant and antimicrobial properties of Moltkia petraea (Tratt.) Griseb. flower, leaf and stem infusions. *Food and Chemical Toxicology,* 48(6): 1537-1542.
- Salopek-Sondi, B.; Piljac-Žegarac, J.; Magnus, V.; Kopjar, N. (2010) Free radical scavenging activity and DNA damaging potential of auxins IAA and 2-methyl-IAA evaluated in human neutrophils by the alkaline comet assay. *Journal of Biochemical and Molecular Toxicology,* 24(3): 165-173.
- Piljac-Žegarac, J. *; Valek, L.; Stipčević, T.; Martinez, S. (2010) Electrochemical determination of antioxidant capacity of fruit tea infusions. *Food Chemistry,* 121(3): 820-825.
- Šamec, D.; Gruz, J.; Strnad, M.; Kremer, D.; Kosalec, I.; Jurišić Grubešić, R.; Karlović, R., Lučić, A. ; Piljac-Žegarac, J. * (2010) Antioxidant and antimicrobial properties of *Teucrium arduini* L. (*Lamiaceae*) flower and leaf infusions. *Food and Chemical Toxicology* 48 (1): 113-119.

- Lucić Vrdoljak, A.; Berend, S.; Želježić, D.; Piljac-Žegarac, J.; Pleština, S.; Kuča, K; Radić, B.; Mladinić, M.; Kopjar, N. (2009) Irinotecan side effects relieved by the use of HI-6 oxime: *in vivo* experimental approach. *Basic & Clinical Pharmacology and Toxicology* 105(6): 401-409.
- Piljac-Žegarac, J. *; Belščak, A.; Piljac A. (2009) Antioxidant capacity and polyphenolic content of blueberry (Vaccinium corymbosum L.) leaf infusions. *Journal of Medicinal Food* 12(3): 1-7.
- Piljac-Žegarac, J. *; Stipčević, T.; Belščak, A. (2009) Antioxidant properties and phenolic content of different floral origin honeys. *Journal of ApiProduct and ApiMedical Science* 1(2): 43-50.
- Piljac-Žegarac, J. *; Valek, L.; Martinez, S.; Belščak, A. (2009) Fluctuations in the phenolic content and antioxidant capacity of dark fruit juices in refrigerated storage. *Food Chemistry* 113: 394-400.
- Piljac, A.; Stipčević, T.; Piljac-Žegarac, J.; Piljac, G. (2008). Successful treatment of chronic decubitus ulcer with 0.1% dirhamnolipid ointment. *Journal of Cutaneous Medicine and Surgery* 12(3): 142-146.

In: Chemistry Researcher Biographical Sketches
Editors: E. S. Huseynov and V. Babayev

ISBN: 978-1-62100-771-5

Chapter 100

BIOGRAPHICAL SKETCH

Name:	Title:	Date of Birth:
Zheng-Wen Li	Temasek laboratories, National University of Singapore	March 12 1950

EDUCATION:

Institution and Location	Degree	Year Conferred	Scientific Field
Shandong University, PRC	B.Sc	1977	
Beijing University of Iron & Steel Technology, PRC	MEng	1982	
University of Manitoba, Canada	PhD	1997	

CONTACT POINTS:

Temasek laboratories, National University of Singapore,
5A, Engineering Drive 1, #07-11, Singapore 117411

RESEARCH AND PROFESSIONAL EXPERIENCE:

Magnetism and magnetic materials, including electromagnetic materials, microwave ferrite materials and devices, ferrite and metallic magnetic films, magnetic nanoparticles, ferrite soft and hard materials, recording materials, NdFeB and rare-earth nitride/ carbide permanent-magnetic alloys, GMR (giant magneto-resistance) and CMR (colossal magneto-resistance) materials.

Professional Appointments:

- Engineer, Beijing Post-Telecommunication Factory, PRC, 1977-1979
- Lecturer Beijing University of Science & Tech, PRC, 1981-1989
- Visiting Scholar, University of Manitoba, Canada, 1989-1993
- Post Doc. University of Manitoba, Canada, 1991-1999
- Visiting Scientist, National University of Singapore, Singapore, 1999-2000
- Senior Research Scientist, Temasek Laboratories, National University of Singapore, Singapore, 2001-present

Publications during last three years:

Z. W. Li, Z. H. Yang, and L. B. Kong, “Resonancelike dispersion for W-type barium ferrite composites with c-axis anisotropy”, *JOURNAL OF APPLIED PHYSICS*, 109 (2011): 0633916

Z. W. Li, L B Kong and Z YANG, "Ultrabroad bandwidth of single-layer electromagnetic attenuation". *APPLIED PHYSICS LETTERS*, 96 (2010): 092507.

Z. W. Li, Z. H. Yang, and L. B. Kong, “Ultrabroad bandwidth and matching characteristics for spinel ferrite composites with flaky fillers”, *JOURNAL OF APPLIED PHYSICS*, 108 (2010): 063927

Kong, L B, Z. W. Li, Z YANG and R HUANG, "Greatly Enhanced Permeability and Expanded Bandwidth for Spinel Ferrite Composites with Flaky Fillers". *IEEE TRANSACTIONS ON MICROWAVE THEORY AND TECHNIQUES*, (2010). (United States).

YANG*, Z, Z. W. Li, and L B Kong*, "One-step synthesis of Ni0.23Cu0.11Zn0.66Fe2O4 ferrite nanowire arrays using a template method". *JOURNAL OF ALLOYS AND COMPOUNDS*, 501, no. 1 (2010): 173-176.

YANG*, Z, Z. W. Li, and L B Kong*, "Structural and magnetic properties of free-standing Ni0.23Cu0.11Zn0.66Fe2O4 thick films prepared using a modified tape-casting method". *JOURNAL OF MAGNETISM AND MAGNETIC MATERIALS*, 322, no. 5 (2010): 557-561.

Z. W. Li, R HUANG and L B Kong, "Greatly enhanced azimuthal permeability of a ferrite core with a wire coil metamaterial". *APPLIED PHYSICS LETTERS*, 94 (2009): 162502.

Z. W. Li, R HUANG and L B Kong, "Permeability and resonance characteristics of metamaterial constructed". *JOURNAL OF APPLIED PHYSICS*, 106 (2009): 103929. 2.

Z. W. Li, L B Kong and Z YANG, "Study of matching characteristics for Ni0.97-xZnxCo0.03Fe2O4 spinel ferrites". *JOURNAL OF APPLIED PHYSICS*, 105 (2009): 113912. (United States).

Z. W. Li, G.Q. Lin, Y.P. WU and L B Kong, "Doping effect on microwave reflection characteristics for W-type ferrite/epoxy composites". *JOURNAL OF PHYSICS D-APPLIED PHYSICS*, 42 (2009): 95007.

Z. W. Li, G.Q. Lin, Y.P. WU and L B Kong, "High-frequency properties and attenuation characteristics of WBa hexaferrite composites with doping of various oxides". *IEEE TRANSACTIONS ON MAGNETICS*, 45, no. 2 (2009): 670-677. (United States).

HUANG*, R, Z. W. Li, L B Kong, L Liu and S Matitsine, "ANALYSIS AND DESIGN OF AN ULTRA-THIN METAMATERIAL ABSORBER". *Progress in Electromagnetics Research*, PIERB 14 (2009): 407-429.

Z. W. Li, G.Q. Lin and L B Kong, "Microwave reflection characteristics of Co2Z barium ferrite composites with various volume concentration". *IEEE TRANSACTIONS ON MAGNETICS*, 44, no. 10 (2008): 2255-2261.

Z. W. Li, Yeow Beng Gan, Xin Xu and G.Q. Lin, "Characteristics of effective permeability and resonance frequency for barium ferrite/epoxy composites". *JOURNAL OF APPLIED PHYSICS*, 103 (2008): 073901.

Z. W. Li, G.Q. Lin, Y.P. WU and L B Kong, "High-frequency magnetic properties and attenuation characteristics for barium ferrite composites". *JOURNAL OF MAGNETISM AND MAGNETIC MATERIALS*, 321 (2008): 734-737.

Kong*, L B, M. L. S. Teo*, Z. W. Li, G.Q. Lin and Y B Gan, "Development of magneto-dielectric materials based on Li-ferrite ceramics: III. Complex relative permeability and magneto-dielectric properties". *JOURNAL OF ALLOYS AND COMPOUNDS*, 459, no. 1-2(2008): 567-582.

Teo*, M. L. S., L B Kong*, Z. W. Li, G.Q. Lin and Y B Gan, "Development of magneto-dielectric materials based on Li-ferrite ceramics: II. DC resistivity and complex relative permittivity". *JOURNAL OF ALLOYS AND COMPOUNDS*, 459, no. 1-2 (2008): 567-575.

Teo*, M. L. S., L B Kong*, Z. W. Li, G.Q. Lin and Y B Gan, "Development of magneto-dielectric materials based on Li-ferrite ceramics: I. Densification behavior and microstructure development". *JOURNAL OF ALLOYS AND COMPOUNDS*, 459, no. 1-2 (2008): 557-566.

In: Chemistry Researcher Biographical Sketches
Editors: E. S. Huseynov and V. Babayev
ISBN: 978-1-62100-771-5

Chapter 101

BIOGRAPHICAL SKETCH

Name:	Title:	Date of Birth:
Jun Zhu	National center for birth defects monitoring of China	Jan.24, 1964

EDUCATION:

Institution and Location	Degree	Year Conferred	Scientific Field
Department of Embry, Chongqing Medical University.	M.D	1985~1988	
Department of pediatrics, Chongqing Medical University,.	BS	1980~1985	
Research maternal and children health information in Japanese national institute for women		2001	
Visiting scholars, on epidemiology and medical genetics in Canadian Albert university		2004	

CONTACT POINTS:

17, Sec 3, RenMin Nanlu, Chengdu, China 610041

RESEARCH AND PROFESSIONAL EXPERIENCE:

1996- present £ºNational Project of Maternal and Child Health Surveillance£¨including national surveillance on maternal mortality ratio, national surveillance on child mortality under 5 year old and national surveillance on birth defects£©£¨Ministry of Health of China, PI£¬£¤7.5 millions£©

2005-2010£ºDecreasing MMR and neo Reducing Maternal Mortality and eliminating Newborn Tetanus (Ministry of Health of China, PI£¬£¤1.0 millions£©

2011-2012£ºEvaluation on project of folic acid supplementation on women with reproductive age among the measurements of Chinese medical reform (Ministry of Health of China and UNICEF , PI£¬£¤0.5 millions£©

2008-2009: Improving China's information system for maternal and child health surveillance(WP/2008/CHN/01) £¨UNICEF, PI£¬£¤4.8 millions£©

2006-2010£º The further construction of national surveillance system for £¨WHO£¬ PI£¬£¤0.68 millions£©

2008-2009: Study of evaluation on earthquake impact on maternal and child health(MCH) information system£¨WHO£¬ PI£¬£¤0.5 millions£©

2010-2013: Research on environmental and genetic factors interaction on congenital heart disease (Ministry of Science and Technology of China, PI, £¤2.8 millions)

2008-2010£ºThe effect evaluation of comprehensive measurements of birth defects intervention in China (Ministry of Science and Technology of China, PI, £¤5.3 millions)

2003-2007 Establish Chinese database and DNA bank for Pedigrees of Birth Defects (CMB£¬Co-PI, US$373,500)

Professional Appointments:

1996- present: Execute Director of National Center for Birth Defects Monitoring of China

Professor of maternal and children health, public health

Publications during last three years

1. Mortality rate for children under 5 years of age in China from 1996 to 2006. Public Health (2011, received, published by May 2011)
2. Preventable maternal mortality: Geographic/rural¨Curban differences and associated factors from the population-based maternal mortality surveillance system in China. BMC public health. 2011£»11:243 doi:10.1186/1471-2458-11-243
3. Assessing the Trend of Gastroschisis Prevalence in China from 1996 to 2007 using Two Analytical Methods. Birth Defects Research (Part A), 2011, 91:177-184
4. Trends in maternal mortality due to obstetric hemorrhage in urban and rural China, 1996¨C2005. J. Perinat. Med. 2011, 39:35-41

5. Maternal mortality in China, 1996-2005. Int J Gynaecol Obstet. 2010; 110(2):93-6.
6. A study on rural-urban differences in neonatal mortality rate in China, 1996-2006. J Epidemiol Community Health. 2010; 64:935-936
7. Low folic acid supplement intake rate among women in northern China with a high-prevalence of neural tube defects, 2008. Preventive Medicine, 2010; 51:338-339
8. Time trends in oral clefts in chinese newborns: data from the chinese national birth defects monitoring network. Birth defects research, 2010 Jan; 88(1):41-7.
9. Folic acid awareness and intake among women in areas with high prevalence of neural tube defects in China: A cross-sectional study. Public health nutrition, 2011, Feb 16:1-7
10. The profile of newborn screening coverage in China. J Med Screen 2009; 16:163-166.
11. Spatial distribution patterns of anorectal atresia/stenosis in China: Use of two-dimensional graph-theoretical clustering, world gastroenterol; 2009-15(22):2787-93.

In: Chemistry Researcher Biographical Sketches
Editors: E. S. Huseynov and V. Babayev

ISBN: 978-1-62100-771-5

Chapter 102

BIOGRAPHICAL SKETCH

Name:	Title:	Date of Birth:
Sergei P.Zhvavyi	B.I. Stepanov Institute of Physics of NAS of Belarus, Minsk, Belarus	1947

EDUCATION:

Institution and Location	Degree	Year Conferred	Scientific Field
Belorussian State University		1971	

CONTACT POINTS:

68 Prospekt Nezavisimosti, Minsk BY-220072, Republic of Belarus

RESEARCH AND PROFESSIONAL EXPERIENCE:

He has authored and co-authored over 110 papers in journals and conference proceedings. His current research interest include: micro-, nanostructures based on anodic alumina, properties of anodic alumina oxide, interaction of nanosecond laser irradiation with semiconductors and dielectrics.

Professional Appointments:

Ph.D. and Dr.Sc. in Physics at Belarus Academy of Sciences, in 1978 and 2006, respectively.

Since 2007 he is chief researcher B.I. Stepanov Institute of Physics of the National Academy of Sciences of Belarus.

Publications during last three years:

1. N. I. Mukhurov, S. P. Zhvavyi, I. V. Gasenkova, S. N. Terekhov, P. P. Pershukevich, V. A. Orlovich / Photoluminescence of F-centers in films of anodic alumina // Journal of Applied Spectroscopy. - 2010. - V. 77, N. 4. - P.549-555.
2. S.N. Terekhov, P. Mojzes, S.M. Kachan, N.I. Mukhurov, S.P. Zhvavyi, A.Yu. Panarin, I.A. Khodasevich, V.A. Orlovich, A. Thorel, F. Grillon P.-Y. Turpin. / A comparative study of surface enhanced Raman scattering from silver-coated anodic aluminum oxide and porous silicon // Journal of Raman Spectroscopy. - 2011. - Vol. 42, Issue 1. - P. 12-20.
3. N.I. Mukhurov, G.I. Efremov, S.P. Zhvavyi. / Embodiments of electrostatic torsion microscaners of optimized functional parameters. // Nano- and microsystemnaya technique (in Russia). - 2010. - š 1. - Ñ.7-15.
4. N.I. Mukhurov, G.I. Efremov, S.P. Zhvavyi. / Theoretical modeling plane-parallel two-electrode microactuators. // Nano- and microsystemnaya technique (in Russia). - 2011. - š1. - C.15-23.
5. N.I. Mukhurov, G.I. Efremov, S.P. Zhvavyi. / A functional capabilities the electrocurrent microrelay // Nano- and microsystemnaya technique (in Russia). - 2011. - š 3. - C.39-42.
6. S.N. Terekhov A.Yu. Panarin N.I. Mukhurov S.P. Zhvavyi I.A. Khodasevich V.A. Orlovich P. Mojzes. / Fabrication of silver nanostructures over anodic aluminum oxide for efficient surface-enhanced Raman scattering. // In "Physics, Chemistry and Applications of Nanostructures. Reviews and Short Notes to Nanomeeting-2009" (V.I. Borisenko, S. V. Gaponenko, V. S. Gurin, Eds.), World Scientific Publishing Co., New Jersey, London, Singapore, Bejing, Shanghai, Hong-Kong, Taipei, Chennai. (2009) p. 503-506.
7. Mukhurov N.I., Zhvavyi S.P., Gasenkova I.V., Terekhov S.N., Panarin A.J., Orlovich V.A., Zhavnerko G.K / Study of photoluminescent properties of anodic aluminum oxide // Proceedings of 17th Intern. Symposium "Nanostructures: Physics and Technology" (NANO2009) Minsk, Belarus, june 22-26, 2009. P.210-211.
8. N.I.Mukhurov, S.S.Krivonosov, I.V.Gasenkova, S.P.Zhvavyi, E,V.Ostapenko / Device for electrochemical oxidation of aluminium with combined influence // Proceedings of 11 international conference and seminar on micro/nanotechnologies and electron devices EDM'2010, Erlagol, Altai, June 30 - July 4, 2010. Novosibirsk, Russia, 2010. - P.278-280.

In: Chemistry Researcher Biographical Sketches
Editors: E. S. Huseynov and V. Babayev
ISBN: 978-1-62100-771-5

Chapter 103

BIOGRAPHICAL SKETCH

Name:	Title:	Date of Birth:
Evangelos Zoidis	Dr	April 20th 1970

EDUCATION:

Institution and Location	Degree	Year Conferred	Scientific Field
Department of Nutritional Physiology and Feeding, Faculty of Animal Science and Aquaculture, Agricultural University of Athens			
Dipl. nat. sc. (Biochemistry-Molecular Biology)	PhD		Biochemistry

CONTACT POINTS:

Department of Nutritional Physiology and Feeding, Faculty of Animal Science and Aquaculture, Agricultural University of Athens, 75 Iera Odos, 11855, Athens, Greece

RESEARCH AND PROFESSIONAL EXPERIENCE:

Evangelos Zoidis graduated (1996) with a Diploma in Biochemistry from the Faculty of Sciences, Biochemistry Department, of the University of Zurich (Switzerland). In 2000 he attained a PhD in Natural Sciences from the University of Zurich. He then spent 2½ years (2000-2003) as post-doctoral fellow at the University Hospital of Zurich. In 2004 he joined the Agricultural University of Athens (Greece) where he worked 3 years as senior scientist and was then (2008) appointed as Lecturer at the Department of Nutritional Physiology and Feeding, Faculty of Animal Sciences and Aquaculture

The research interests focus on nutritional physiology, endocrinology and molecular metabolism:

- Study of phosphate, sodium and vitamin D metabolism and genetic mutations that affect them
- Relationship between food interactions and physiology/homeostasis of the skeletal and digestive system
- Sodium-phosphate transport systems and glucose transport systems in vivo and in vitro
- Study of the pituitary-endocrine axis pancreas: growth hormone and actions of IGF I
- Gene/protein expression and cloning
- Endopeptidases/membrane isoenzymes and control of gene expression
- Selenium/selenoproteins: metabolism and actions/gene expression

Professional Appointments:

2008-present: Lecturer at the Agricultural University of Athens
2004 - 2008: Senior scientist at the Department of Nutritional Physiology and Feeding, Faculty of Animal Sciences and Aquaculture, Agricultural University of Athens
2000 - 2003**:** Postdoctoral researcher (fellow of the Swiss National Research Foundation, SNF) in the Department of Metabolism, University Hospital of Zurich, Switzerland
2000: Doctor of Philosophy (Ph.D.), Faculty of Sciences, Biochemistry Department, University of Zurich, Switzerland (fellow of the National Research Foundation, SNF)
1996: Diploma in Biochemistry (Faculty of Sciences, Biochemistry Department, University of Zurich, Switzerland)

Awards and Honors:

April 1997: Swiss National Science Foundation, PhD funding, grant no. 32-46808.97 (duration 40 months)
June 2000: Swiss National Science Foundation, Postdoctoral Research funding (duration 24 months)
March 2001: Pharmacia Leonardo Award 2000, from the Italian Research Organization, Milan, Italy
November 2004: PHOENIX Pharmacy-Research Award 2004, for scientific work in the field of Biochemical-Pharmaceutical Technology

Publications during last three years:

- Pappas A.C., Zoidis E., Surai P.F., Zervas G. (2008). Selenoproteins and maternal nutrition. Comp. Bioch. Physiol. B.-151, 361-372

- Pappas A.C., Zoidis E., Fegeros K., Surai P.F., Zervas G. (2009). Relation of cadmium to other elements and the antioxidant system. Nova Science Publishers, Inc., 400 Oser Avenue, Suite 1600, Hauppauge, NY 11788
- Poulopoulou I., Hatzigeorgiou I., Zoidis E., Tsoufi V., Masouras T. (2009). Effects of terpenes administration in the physiochemical characteristics of goat milk. Greek J. Dairy Sci. Technol., Special Issue 2009, 72-81
- Pappas A.C., Zoidis E., Fegeros K., Surai P.F., Zervas G. (2010). Cadmium toxicity and the antioxidant system. Nova Science Publishers, Inc., 400 Oser Avenue, Suite 1600, Hauppauge, NY 11788
- Jin C., Zoidis E., Ghirlanda C., Schmid Ch. (2010). Dexamethasone and Cyclic AMP Regulate Sodium Phosphate Cotransporter (NaPi-IIb and Pit-1) mRNA and Phosphate Uptake in Rat Alveolar Type II Epithelial Cells. Lung 188, 51-61
- Zoidis E., Pappas A.C., Georgiou C.A., Komaitis ., Fegeros K. (2010). Selenium affects the expression of GPx4 and catalase in the liver of chicken. Comp. Bioch. Physiol. B.-155, 294-300
- Pappas A.C., Zoidis E., Theophilou N., Zervas G., Fegeros K. (2010). Effects of paligorskite on broiler performance, feed technological characteristics and litter quality. Appl. Clay Sci. 49, 276-280
- Papadomichelakis G., Mountzouris K.C., Zoidis E., Fegeros K. (2011). Influence of dietary benzoic acid addition on nutrient digestibility and selected biochemical parameters in fattening rabbits. Anim. Feed Sci. Technol. 163, 207-213
- Zoidis E., Ghirlanda-Keller C., Schmid Ch. (2011). Stimulation of glucose transport in osteoblastic cells by parathyroid hormone and insulin-like growth factor I. Mol. Cell. Biochem. 348, 33-42
- Pappas A.C., Zoidis E., Georgiou C.A., Demiris N., Surai P.F., Fegeros K. (2011). Influence of organic selenium supplementation on the accumulation of toxic and essential trace elements involved in the antioxidant system of chicken. Food Addit. Contam.-A. 28, 446-454
- Poulopoulou I., Zoidis E., Massouras T., Hadjigeorgiou I. (2011). Terpenes transfer to milk and cheese after oral administration to sheep fed indoors. J. Anim. Physiol. Anim. Nutr. (in press)
- Pappas A.C., Zoidis E., Papadomichelakis G., Fegeros K. (2011). Supranutritional selenium level affects fatty acid composition and oxidative stability of chicken breast muscle tissue. J. Anim. Physiol. Anim. Nutr. (in press)

In: Chemistry Researcher Biographical Sketches
Editors: E. S. Huseynov and V. Babayev

ISBN: 978-1-62100-771-5

Chapter 104

BIOGRAPHICAL SKETCH

Name:	Title:	Date of Birth:
Zou Jin-Long	PhD	1980-04-11

EDUCATION:

Institution and Location	Degree	Year Conferred	Scientific Field
Key Laboratory of Functional Inorganic Materials Chemistry, Ministry of Education, School of Chemistry and Materials Science, Heilongjiang University, Harbin 150080, China.			

CONTACT POINTS:

School of Chemistry and Materials Science, Heilongjiang University, 74 Xuefu Road, Nangang District, 150080, Harbin, Heilongjiang Province, China.

RESEARCH AND PROFESSIONAL EXPERIENCE:

2003-2008, Harbin institute of technology, PhD student;
2008-2010, Heilongjiang University, Lecturer;
2011-, Heilongjiang University, Associate Professor.

Professional Appointments:

Associate Professor

Publications during last three years:

1. Zou Jinlong*, Yu Xiujuan, Dai Ying and Xu Guoren. Utilization of Water and Wastewater Sludge for Production of Lightweight-Stabilized Ceramsite, as a contributing chapter for book <Fluid Waste Disposal>, 2009, Nova Science Publishers. USA.
2. Guoren Xu, Zou Jinlong* and Guibai Li. Stabilization/ Solidification of Heavy Metals in Sludge Ceramsite and Leachability Affected by Oxide Substances. Environmental Science and Technology. 2009, 43(15): 5902-5907
3. Guoren Xu, Zou Jinlong* and Guibai Li. Ceramsite Made with Water and Wastewater Sludge and its Characteristics Affected by SiO_2 and Al_2O_3. Environmental Science and Technology. 2008, 42(19): 7417-7423
4. G. R. Xu, J. L. Zou* and G. B. Li. Ceramsite Obtained from Water and Wastewater Sludge and its Characteristics Affected by (Fe_2O_3+CaO+MgO)/ (SiO_2+Al_2O_3). Water Research. 2009, 43(11): 2885-2893
5. G. R. Xu, J. L. Zou* and G. B. Li. Stabilization of Heavy Metals in Ceramsite Made from Water and Wastewater Sludge. Water Research. 2010, 44(9): 2930-2938
6. J. L. Zou*, Y. Dai, T. H. Sun, Y. H. Li, G. B. Li and Q. Y. Li. Effect of Amended Soil and Hydraulic Load on Enhanced Biological Nitrogen Removal in SWIS. Journal of Hazardous Materials. 2009, 163(2-3): 816-823
7. J. L. Zou*, G. R. Xu and G. B. Li. Ceramsite Obtained from Water and Wastewater Sludge and its Characteristics Affected by Fe_2O_3, CaO, and MgO. Journal of Hazardous Materials. 2009, 165(1-3): 995-1001.
8. G. R. Xu, J. L. Zou* and G. B. Li. Effect of sintering temperature on the characteristics of sludge ceramsite. Journal of Hazardous Materials. 2008, 150(2): 394-400.
9. G. R. Xu, J. L. Zou* and G. B. Li. Stabilization of heavy metals in ceramsite made with sewage sludge. Journal of Hazardous Materials. 2008, 152(1): 56-61.
10. G. R. Xu, J. L. Zou* and G. B. Li. Solidification and Leaching Behaviors of Cr^{6+} in Sludge Ceramsite. Journal of Hazardous Materials. 2008, 153(3): 1031-1035.
11. G. R. Xu, J. L. Zou* and Y. Dai. Utilization of dried sludge for making ceramsite. Water Science and Technology, 2006, 54(9): 69-79.
12. Zou Jinlong*, Dai Ying. Leaching Characteristics of Heavy Metals and Utilization of Filter Media in BAF. Environmental science. 2007, 28(10): 231-235.

PART 2 - RESEARCH SUMMARIES IN CHEMISTRY

In: Chemistry Researcher Biographical Sketches
Editor: E. S. Huseynov and V. Babayev
ISBN: 978-1-62100-771-5

Chapter 105

ADVANCES IN ADSORPTION TECHNOLOGY

Bidyut Baran[1] and Kim Choon Ng[2]
[1]Saha Kyushu University, Japan
[2]National University of Singapore, Singapore

RESEARCH SUMMARY

The efficient utilization of energy at low heat source temperature is a key issue for both industry and academia. In recent years, the heat-activated adsorption cycles are proven to be practical and energy efficient method for converting low-temperature waste heat into useful effects such as cooling, freezing, heating, storage and desalination and this is reflected by the dearth of literature publications. The adsorption technology would steer towards the key development of heat activated machines for the energy industry and they could offer cost-effective solutions for achieving energy efficiency and environment sustainability. This edited book covers the state-of-the art of adsorption research and technologies for relevant applications with the objectives of energy efficiency and sustainability. It consists of fifteen chapters from renowned experts in adsorption fields around the world covering the aspects of adsorption fundamentals, adsorbent materials characterizations, adsorption cooling, storage, and heat pump systems. This book examines new research directions in this frontier field.

In: Chemistry Researcher Biographical Sketches
Editor: E. S. Huseynov and V. Babayev
ISBN: 978-1-62100-771-5

Chapter 106

ADVANCES IN CHEMICAL MODELING

Mihai V. Putz
West University of Timisoara, Timisoara, Romania

RESEARCH SUMMARY

This new book brings together the theoretical, mathematical, quantum and computational methodologies of approaching structural, bonding and reactivity characterizations of the chemical systems.

In: Chemistry Researcher Biographical Sketches
Editor: E. S. Huseynov and V. Babayev

ISBN: 978-1-62100-771-5

Chapter 107

ANALYTICAL CHEMISTRY OF CADMIUM: SAMPLE PRE-TREATMENT AND DETERMINATION METHODS

Antonio Moreda-Piñeiro[1] and Jorge Moreda-Piñeiro[2]
[1]Department of Analytical Chemistry, University of Santiago de Compostela, Spain
[2]Department of Analytical Chemistry. University of A Coruña,
Campus da Zapateira, Spain

RESEARCH SUMMARY

In this book, a brief description of sampling strategies, preservation and handling of environmental samples (liquids, solids and air) are commented on, with particular emphasis on the subsequent cadmium determination. The main analytical techniques for cadmium determination are briefly described, and they are compared in base on their advantages and drawbacks. In addition, the use of automatic analyzers for environmental monitoring are also described, as well as the main sample pre-treatment methods for liquid and solid environmental samples. Well-established methodologies and current trends are reviewed for cadmium determination in environmental matrices such as water, soils, sediments, biota, and atmospheric particulate matter. In addition, cadmium speciation, mainly cadmium fractionation in soil, sediment, sludge and atmospheric particulate matter (BCR method), and isolation and separation of cadmium bound to metallothioneis in biota is reviewed. Finally, a source of literature on standard and official methods for cadmium determination in environmental materials is provided in this book.

In: Chemistry Researcher Biographical Sketches
Editor: E. S. Huseynov and V. Babayev
ISBN: 978-1-62100-771-5

Chapter 108

ARTIFICIAL NEURAL NETWORK MODELING OF WATER AND WASTEWATER TREATMENT PROCESSES

Ali R. Khataee and Masoud B. Kasiri
University of Tabriz, Tabriz, Iran

RESEARCH SUMMARY

Artificial neural networks (ANNs) are computer based systems that are designed to simulate the learning process of neurons in the human brain. ANNs have been attracting great interest during the last decade as predictive models and pattern recognition. Artificial neural networks possess the ability to "learn" from a set of experimental data without actual knowledge of the physical and chemical laws that govern the system. Therefore, ANNs application in data treatment is high, especially where systems present nonlinearities and complex behavior. This book describes the application of artificial neural networks for modeling of water and wastewater treatment processes.

In: Chemistry Researcher Biographical Sketches
Editor: E. S. Huseynov and V. Babayev

ISBN: 978-1-62100-771-5

Chapter 109

BIOMEDICAL APPLICATIONS OF LANTHANUM

Harriet Nilsson[1], Anca Dragomir[1] and Godfried M. Roomans[2]
[1] University of Uppsala, Uppsala, Sweden
[2] University of Orebro.Orebro, Sweden

RESEARCH SUMMARY

Lanthanum has been used as a tracer to study tight junctions in epithelia by transmission electron microscopy. The key function of the tight junction complex (TJC) is to limit the paracellular permeability by forming a selectively permeable barrier between the apical and basolateral compartment of the extracellular space. Impaired regulation of the TJC causes a variety of diseases, such as autoimmune diseases, cancer development, infections, allergies and asthma. The advantage of lanthanum as an opaque probe is its small ionic radius (0.11 nm), compared to alternative methods, such as radioactive isotope techniques using extracellular tracers and light microscopy techniques with larger polar molecules. This book presents a variety of biomedical applications of lanthanum.

In: Chemistry Researcher Biographical Sketches
Editors: E. S. Huseynov and V. Babayev

ISBN: 978-1-62100-771-5

Chapter 110

BORON HYDRIDES, HIGH POTENTIAL HYDROGEN STORAGE MATERIALS

Umit B. Demirci and Philippe Miele
Universite Claude Bernard Lyon, Villeurbanne, France

RESEARCH SUMMARY

Boron hydrides are hydrogen storage materials which are the object of intensive investigation because they pose tangible solution to the hydrogen storage issue. Today many boron hydrides are intensively investigated: lithium borohydride LiBH4, sodium borohydride NaBH4, potassium borohydride KBH4, new borohydrides like e.g. Ca(BH4)2, Zn(BH4)2 or Zr(BH4)4, combination of hydrides like e.g. LiBH4 and LiNH2 or LiBH4 and MgH2, ammoniaborane NH3BH3, alkali amidoboranes like e.g. LiNH2BH3 or NaNH2BH3, and so on. This new book reviews research on boron hydrides and gives a general view of the perspectives of application.

In: Chemistry Researcher Biographical Sketches
Editors: E. S. Huseynov and V. Babayev
ISBN: 978-1-62100-771-5

Chapter 111

CHEMISTRY AND BIOCHEMISTRY: FROM PURE TO APPLIED SCIENCE, NEW HORIZONS-VOL. 3 (A FESTSCHRIFT IN HONOR OF THE 75TH BIRTHDAY OF PROFESSOR GENNADY E. ZAIKOV)

***Eli M. Pearce*[1] *and G. E. Zaikov*[2]**

[1] Herman Mark Polymer Research Inst., Brooklyn Polytechnic Univ. Brooklyn, US
[2] N.M Emanuel Inst. of Biochemical Physics, Russian Academy of Sciences, Moscow, Russia

RESEARCH SUMMARY

The present book is the third book devoted to 75 anniversary of G.E. Zaikov's birth and 55 years of his creative activity.

This new volume contains information about the plasticity structural analysis of polymer nanocomposites, filled by carbon nanotubes, grafting of some bio-fibres with carboxylic acids under cold plasma conditions, achievements in textile dyes removal using adsorption, a reactive medium viscosity change in cross-linking process of epoxy polymers, the side groups influence on polymers limiting mechanical characteristics.

In: Chemistry Researcher Biographical Sketches
Editors: E. S. Huseynov and V. Babayev

ISBN: 978-1-62100-771-5

Chapter 112

CHEMISTRY AND CHEMICAL ENGINEERING RESEARCH PROGRESS

A. K. Haghi
University of Guilan, Iran

RESEARCH SUMMARY

Chemical engineering is the branch of engineering that deals with the application of physical science, and life sciences with mathematics, to the process of converting raw materials or chemicals into more useful or valuable forms. In addition to producing useful materials, modern chemical engineering is also concerned with pioneering valuable new materials and techniques - such as nanotechnology, fuel cells and biomedical engineering. This new book reviews research related to chemical industry, chemical engineering and other connected areas.

In: Chemistry Researcher Biographical Sketches
Editors: E. S. Huseynov and V. Babayev
ISBN: 978-1-62100-771-5

Chapter 113

CHITOSAN-BASED HYDROGELS FOR TISSUE ENGINEERING APPLICATIONS

G. Luna-Bárcenas, E. Prokhorov, E. Elizalde-Peña, A. Nuno-Licona, I. C. Sanchez, J. E. Gough, C. Velasquillo-Martinez and C. E. Schmidt
Nacional Libramiento Norponiente, Querétaro, México

RESEARCH SUMMARY

The polyelectrolyte complexes formed due to electrostatic attraction between the polymers, have improved properties when compared to hybrid CTS-g-GMA, which provides a wide range of applications in the biomedical field. All materials have been characterized by different analytical techniques such as infrared spectroscopy (FTIR), X-ray diffraction (XRD), thermal analysis (DSC and TGA), and the results were compared to the precursor materials (chitosan, X, HA, and CTS-g-GMA). Due to the HA nature, the film obtained from this reaction has been assessed for use as a patch for wound healing; whereas the properties showed by the hydrogels obtained from the reaction with X, make them very promising for applications in the treatment of recovery of spinal cord injuries. This book presents results which strongly support the future use of hydrogels as scaffolds for tissue engineering and recovering.

In: Chemistry Researcher Biographical Sketches
Editors: E. S. Huseynov and V. Babayev
ISBN: 978-1-62100-771-5

Chapter 114

Cyclic B-Ketoesters: Synthesis and Reactions

M. A. Metwally and E. G. Sadek
Department of Chemistry, Faculty of Science, University of Mansoura, Egypt

Research Summary

The chemistry of cyclic b-ketoesters has received considerable attention because they serve as highly important intermediates in many organic synthesis. The wide application of these compounds in organic synthesis, depends chiefly on the reactivity of the methylene group, and the ready hydrolysis of the ester group, and its substituted derivatives to carboxlic acids and ketones. The syntheses and reactions of cyclic b-ketoesters are reviewed in a formal way.

In: Chemistry Researcher Biographical Sketches
Editors: E. S. Huseynov and V. Babayev
ISBN: 978-1-62100-771-5

Chapter 115

DZHEMILEV REACTION IN ORGANIC AND ORGANOMETALLIC SYNTHESIS

Vladimir A. D'yakonov
Institute of PetroChemistry and Catalysis of Russian Academy of Sciences,
Bashkortostan, Russian Federation

RESEARCH SUMMARY

In the last 10−15 years the discovery of catalytic cycloalumination, cyclomagnesiation, and ethylmagnesiation reactions of olefins, dienes and allenes with the aid of alkyl and halogen alkyl Al and Mg derivatives under the action of Ti, Zr and Co complexes to obtain new classes of cyclic and acyclic organometallics namely aluminacyclopropanes, aluminacyclopropenes, aluminacyclopentanes, aluminacyclopent-2-enes, aluminacyclopenta-2,4-dienes, magnesacyclopentanes, magnesacyclopent-2-enes, and magnesacyclopenta-2,4-dienes are among the sound achievements in the field of organic and organometallic synthesis. These reactions are considered in the world scientific literature as Dzhemilev reactions. The monograph covers the data concerned the revealed first by Professor Dzhemilev phenomenon of catalytic replacement of transition metal atoms in metallacarbocycles by nontransition metal atoms to afford the said classes of organometallic compounds (OMC) and also the history of this discovery. Modern concepts and outlooks on mechanism of these reactions as well as the structure of the key intermediates responsible for the formation of target OMC are considered.

In: Chemistry Researcher Biographical Sketches
Editors: E. S. Huseynov and V. Babayev

ISBN: 978-1-62100-771-5

Chapter 116

ELECTROCHEMICAL HYDROGEN AND METALS: ABSORPTION, DIFFUSION AND EMBRITTLEMENT PREVENTION IN CORROSION AND ELECTROPLATING

Sergei Beloglazov
Immanuel Kant University of Kaliningrad, Kaliningrad, Russia

RESEARCH SUMMARY

Electrochemical hydrogen evolves both in the cathodic process of corrosion on the interface aqueous solution of electrolyte / Me (steel) and in the cathodic processes of Me (Zn, Cd, Ni, Cr, Sn, Cu, etc) electrodeposition for steel corrosion protection. Absorption of the part of cathodic discharged hydrogen deteriorates all mechanical properties of metal and results in hydrogen embrittlement. During the last three decades, many works on research of systems H-Me have been published and the new materials extremely sensitive to hydrogen fragility have appeared. This new and important book reviews research results concerning the influence of organic brightening compounds on steel hydrogen embrittlement at electrodeposition of the above mentioned metals.

In: Chemistry Researcher Biographical Sketches
Editors: E. S. Huseynov and V. Babayev
ISBN: 978-1-62100-771-5

Chapter 117

ELECTROCHEMICAL PROPERTIES AND APPLICATIONS OF IONIC LIQUIDS

Angel A. J. Torriero and Muhammad J. A. Shiddiky
Monash University, Victoria, Australia

RESEARCH SUMMARY

This book presents the latest research in electrochemical properties and applications of ionic liquids. While there is no universally agreed upon definition, an ionic liquid may be conveniently described as a compound composed entirely of ions that is a liquid at temperatures less than 100 °C. However, this is an arbitrary definition employed to distinguish ionic liquids from classically well-known molten salts. This book addresses a comprehensive overview of the area, because it is obvious that ionic liquids have the ability to offer many advantages, but also some disadvantages, over traditional molecular solvent (electrolyte) media in the field of electrochemistry.

In: Chemistry Researcher Biographical Sketches
Editors: E. S. Huseynov and V. Babayev

ISBN: 978-1-62100-771-5

Chapter 118

ELECTROSPRAY TECHNOLOGY FOR THIN-FILM DEPOSITION

Anatol Jaworek
Polish Academy of Sciences, Gdańsk, Poland

RESEARCH SUMMARY

Electrospraying utilizes electrical forces for liquid atomization. Droplets obtained by this method are highly charged, to about half of the Rayleigh limit. The advantage of electrospraying is that the droplets can be extremely small, down to the order of 10's nanometres, and the charge and size of the droplets can be controlled to some extent by electrical means. The application of electrospray technology for thin solid film devices is presented in this book. This book focuses on thin inorganic film deposition for solar cells, fuel cells, lithium batteries, supercapacitors, gas sensors production.

In: Chemistry Researcher Biographical Sketches
Editors: E. S. Huseynov and V. Babayev
ISBN: 978-1-62100-771-5

Chapter 119

EMERGING TOPICS IN ORGANIC CHEMISTRY

Rattaporn Thonggom[1], Thevarak Rochanapruk[2] and Gennady E. Zaikov[3]
[1] Mahidol Univ., Bangkok, Thailand
[2] PTT Phenol Co. Ltd, Chatuchak, Bangkok, Thailand
[3] N.M. Emanuel Institute of Biochemical Physics, Russian Academy of Sciences, Moscow, Russia

RESEARCH SUMMARY

This book is a first attempt of scientific cooperation between Mahidol University and PTT Phenol Company from Thailand and Institute of Biochemical Physics, Russian Academy of Sciences. The authors selected for their cooperation the topics from organic chemistry (quantitative level). Special attention was done for chemical physics and biochemical physics of organic compounds (low- and highmolecular). The authors included some papers about synthesis, investigation of properties and looking for application. The authors expect that the information of this book will be useful for scientists, students and engineers who are working in the field of organic chemistry.

In: Chemistry Researcher Biographical Sketches
Editors: E. S. Huseynov and V. Babayev

ISBN: 978-1-62100-771-5

Chapter 120

HUMIC ACIDS FROM RAW MATERIALS OF THE CZECH REPUBLIC

Libuse Madronova
Research Institute of Inorganic Chemistry, Czech Republic

RESEARCH SUMMARY

In the period of more than twenty years, the Research Institute of Inorganic Chemistry in the Czech Republic, together with various other universities, research institutes and commercial organizations, worked on projects concerned with humic substances. The main reason was that the Czech Republic is rich in deposits of various raw materials potentially suitable for the production of humic substances. Main goals of the investigations were the selection of raw materials, as well as the determination of their chemical and physical properties. This book presents the results of these extensive investigation.

In: Chemistry Researcher Biographical Sketches
Editors: E. S. Huseynov and V. Babayev

ISBN: 978-1-62100-771-5

Chapter 121

INFORMATION ORIGINS OF THE CHEMICAL BOND

Roman F. Nalewajski
Dept. of Theoretical Chemistry, Jagiellonian University, Cracow, Poland

RESEARCH SUMMARY

The Information Theory (IT) is one of the youngest branches of the applied probability theory, in which the probability ideas have been introduced into the field of communication, control, and data processing. It has originated from the needs of practice, to create a theoretical model for the transmission of information, and evolved into an important chapter of the general theory of probability. An understanding of the distribution of information in molecules and its displacements accompanying chemical reactions, which involve the bond-forming and/or bond-breaking processes, is touched on in this book and provides an alternative perspective on molecular electronic structure. An insight into the entropic origins of chemical bonds and their coupling in chemical phenomena is central to many branches of chemistry.

In: Chemistry Researcher Biographical Sketches
Editors: E. S. Huseynov and V. Babayev
ISBN: 978-1-62100-771-5

Chapter 122

LAYERED DOUBLE HYDROXIDES: PRESENT AND FUTURE

Vicente Rives
Universidad de Salamanca, Spain

RESEARCH SUMMARY

Layered double hydroxides are one of the variety of names given to a family of layered materials first discovered in Sweden in 1842. These materials are interesting because their layer cations can be changed among a wide selection, and the interlayer anion can also be (nearly) freely chosen. Like cationic clays, they can be pillared and can exchange interlayer species – thus increasing applications and making new routes to derivatives. The principle areas of application include catalyst support, anion scavengers, polymer stabilizers, and antacids. In the last several years, reviews and studies of LDHs have dealt with these uses. This book aims to update the current body of LDH knowledge from a wide array of views. The first section addresses the synthesis and physiochemical characterization of these materials, and section two focuses on the applications of LDHs.

In: Chemistry Researcher Biographical Sketches
Editors: E. S. Huseynov and V. Babayev

ISBN: 978-1-62100-771-5

Chapter 123

MAGNETIC INTERACTIONS IN OXO-CARBOXYLATE BRIDGED GADOLINIUM (III) COMPLEXES

Laura Cañadillas-Delgado, Óscar Fabelo, Catalina Ruiz-Pérez and Joan Cano
Universidad de La Laguna, Spain

RESEARCH SUMMARY

Magneto-structural studies on polynuclear complexes, aimed at understanding the structural and chemical factors that govern the exchange coupling between paramagnetic centers, are of continuing interest. One of the best illustrative examples corresponds to the di-μ-hydroxo-dicopper(II) complexes where the angle at the hydroxo bridge is the main factor governing the nature and magnitude of the intramolecular magnetic coupling. This book presents a systematic study of complex formation between gadolinium(III) and carboxylate-containing ligands which investigates the factors that govern the magnetic coupling in the digadolinium(III) units.

In: Chemistry Researcher Biographical Sketches
Editors: E. S. Huseynov and V. Babayev

ISBN: 978-1-62100-771-5

Chapter 124

MOLECULAR AND SUPRAMOLECULAR BIOINORGANIC CHEMISTRY: APPLICATIONS IN MEDICAL AND ENVIRONMENTAL SCIENCES. VOLUME 2

***Ana Lucia Ramalho Mercê*[1]*, Judith Felcman*[2]**
***and María Ángeles Lobo Recio*[3]**
[1]Chemistry Dept., Univ. Federal do Paraná, Paraná, Brasil
[2]Pontifícia Univ. Católica do Rio de Janeiro, RJ, Brasil
[3]Saintary Engineering Dept. Univ. Federal de Santa Catarina, SC, Brasil

RESEARCH SUMMARY

This second volume of the series intends to continue helping to fill the gap between research and practicing medical professionals as well as students in general. It will help to bring to anyone interested in the ultimate results and conclusions from research produced among interdisciplinary areas of knowledge, an understanding of the neurodegenerative diseases. It is known today that metal ions participate in the enzymatic chemical reactions of the living bodies either acting as essential elements or being responsible for oxidative damaging processes.

In: Chemistry Researcher Biographical Sketches
Editors: E. S. Huseynov and V. Babayev

ISBN: 978-1-62100-771-5

Chapter 125

MOLECULAR SYMMETRY AND FUZZY SYMMETRY

Xuezhuang Zhao
Department of Chemistry, Nankai University, Tianjin, P.R., China

RESEARCH SUMMARY

Since ancient times, human beings have connected beauty with symmetry. In science, this correlation is found in many areas, from the fundamental laws of physics to the products of modern chemistry. However, it seems emphasize the deadness of perfect symmetry or the dis-symmetry makes the phenomenon. As well known, most molecules will become less symmetrical as substitutions take place. Here, the authors turn to a consideration of how to describe the effect of this kind of imperfect symmetry on molecular properties.

In: Chemistry Researcher Biographical Sketches
Editors: E. S. Huseynov and V. Babayev
ISBN: 978-1-62100-771-5

Chapter 126

NANOTUBE ARRAY SUPERCAPACITOR

Yibing Xie
Southeast University, China

RESEARCH SUMMARY

Nanotube array supercapacitors have been designed for electric double layer and redox capacitance applications using bare titania nanotubes and electroactive nickel oxide or ruthenium oxide modified titania nanotube composite electrodes. Titania nanocomplexes with nanoribbon-nanotube and nanowire-nanotube structure have been synthesized through a two-step anodization process. The well-defined titania has been applied for electric double layer supercapacitor applications. This book presents and discusses current research in the creation and application of nanotube array supercapacitors.

In: Chemistry Researcher Biographical Sketches
Editors: E. S. Huseynov and V. Babayev
ISBN: 978-1-62100-771-5

Chapter 127

Optimization in Polymer Processing

António Gaspar-Cunha and José António Covas
University of Minho, Portugal

Research Summary

Plastics processing is a major industrial activity, which yields components and systems for a wide range of industries, such as packaging, automotive, aeronautics, electrical and electronic, sports and leisure, toys, civil and construction, and agriculture. Most plastic components are manufactured either by extrusion or injection molding, but other techniques such as blow molding and thermoforming are also important. The productivity of these technologies is dictated by the equipment design, choice of the operating conditions and physical properties of the polymer system. This book discusses the recent scientific developments on the optimization of manufacturing engineering problems and applies them to polymer processing technologies.

In: Chemistry Researcher Biographical Sketches
Editors: E. S. Huseynov and V. Babayev
ISBN: 978-1-62100-771-5

Chapter 128

ORGANIC EUROPIUM COMPLEXES AND THEIR APPLICATIONS IN OPTOELECTRONIC DEVICES

Han You and Dongge Ma
Chinese Academy of Sciences, Changchun, P. R. China

RESEARCH SUMMARY

Organic europium complexes are of considerable interest due to their sharp narrow emission spectrum and potentially high emission quantum efficiency, originated from 2D0-7F2 electronic transitions of the central ions. Therefore, organic europium complexes have well applications in organic light-emitting diodes. Furthermore, organic europium complexes also show well electrical switching characteristics in diodes, potential applications as memory devices in information storage. This book reviews the achieved progresses in organic europium complexes and their applications in organic light-emitting diodes and memory devices.

In: Chemistry Researcher Biographical Sketches
Editors: E. S. Huseynov and V. Babayev

ISBN: 978-1-62100-771-5

Chapter 129

PHOSPHAZENES: A WORLDWIDE INSIGHT

***R. De Jaeger*[1] *and M. Gleria*[2]**
[1]Lille University, France
[2]Centro di Stabilità e Reattività dei Composti di Coordinazione, CNR. Padova, Italy

RESEARCH SUMMARY

This book provides the most complete and updated picture of the research presently going on in the field of phosphazene materials all over the world. The large majority of scientists active in the phosphazene domain contributed to this book writing chapters on their own specific activity regarding these materials, also stressing future possible developments of this extremely interesting scientific and technologic area.

Both cyclic and polymeric phosphazenes were described in this book, starting from the most representative and effective synthetic strategies used for their preparation, touching theoretical and spectroscopic research aspects for these materials, to conclude by reporting their possible practical use in several different application domains.

In: Chemistry Researcher Biographical Sketches
Editors: E. S. Huseynov and V. Babayev

ISBN: 978-1-62100-771-5

Chapter 130

PHYSICAL CHEMISTRY EXAMINATIONS

A. O. Ibhadon
Univ. of Hull, Faculty of Science and the Environment, Hull, England

RESEARCH SUMMARY

This book was written to meet the needs and demands of students who find studying physical chemistry a difficult and at times a frustrating experience. The author have observed over several years of teaching this subject at the undergraduate level that students are unrelenting in their desire for so called 'worked examples' and answers to 'typical examination' questions. This book has been prepared primarily to meet this need and assist students at the early years of the undergraduate degree programme in chemistry and chemical engineering understand this subject better and perhaps appreciate the relevance and beauty of the subject as a whole. Worked examples have been provided in every chapter as well as exercises and 'typical examination questions' and solutions to these questions. Care has been taken to ensure that worked examples have been explained in a step by step basis to avoid confusion and misunderstanding on the part of the reader. S.I units have been explained throughout and a reasonably good grounding in mathematics is assumed throughout this book. This book is also aimed at the physical chemistry lecturer/teacher who will discover that students are the same everywhere in their desire for 'worked examples', tutorials and solutions to 'typical examination questions'. This book is thus a useful companion to the physical chemistry lecturer and teacher as well as to his/her students. It is not a replacement for a more comprehensive text on the subject and students of physical chemistry are advised to avail themselves of such texts.

In: Chemistry Researcher Biographical Sketches
Editors: E. S. Huseynov and V. Babayev

ISBN: 978-1-62100-771-5

Chapter 131

POLYMER AND COMPOSITES: THEORY AND PRACTICAL APPLICATIONS

G. V. Kozlov[1], *A. K. Mikitaev*[1] *and G. E. Zaikov*[2]
[1]Kabardino-Balkarian State University, Moscow, Russia
[2]Russian Academy of Sciences, Moscow, Russia

RESEARCH SUMMARY

This book examines polymer nanocomposites filled by inorganic nanofillers. Polymers are considered as nanocomposite matrix only, possessing, as a rule, an invariable structure. In many respects this situation is explained by the absence of a quantitative structural model of polymers amorphous state. This problem becomes particularly important because all structural elements of polymers have sizes of nanometer scale. The development of notions about polymers amorphous state structure within the frameworks of the cluster model of this structure allows to represent amorphous polymer as a quasitwophase system.

In: Chemistry Researcher Biographical Sketches
Editors: E. S. Huseynov and V. Babayev

ISBN: 978-1-62100-771-5

Chapter 132

PREPARATION OF THIN FILM PD MEMBRANES FOR H2 SEPARATION FROM SYNTHESIS GAS AND DETAILED DESIGN OF A PERMEABILITY TESTING UNIT

M. Bientinesi and L. Petarca
Dipartimento di Ingegneria Chimica, Chimica Industriale e Scienza dei Materiali,
Università di Pisa, Pisa, Italy

RESEARCH SUMMARY

Hydrogen is one of the most important chemical products, and it is used in a wide series of industrial fields: chemical, petrochemical, metallurgic, as well as in energy applications. Hydrogen demand in all these fields is continuously growing. Nowadays hydrogen is mainly produced from fossil fuels via processes such as steam reforming, partial oxidation, and gasification. All these processes lead to the production of an H2 containing gas stream (synthesis gas) from which hydrogen has to be separated. Thin-film palladium membranes are one of the most promising technologies for the separation of hydrogen from synthesis gas. In this book, palladium films were deposited above stainless steel porous supports using the electroless plating (ELP) technique. The supports have a 0.1 ìm filter grade. Two different geometry (disc sheet and tubes) were used. Finally, in this book, the detailed design of a permeability testing unit is reported, including the fluid dynamic, thermal and mechanical dimensioning, the selection of materials and equipment, and some safety considerations. The sketches of the two permeation cells (for the two different membrane geometry) are also reported.